D1281430

Probability on Graphs

This introduction to some of the principal models in the theory of disordered systems leads the reader through the basics, to the very edge of contemporary research, with the minimum of technical fuss.

Topics covered include random walk, percolation, self-avoiding walk, interacting particle systems, uniform spanning tree, random graphs, as well as the Ising, Potts, and random-cluster models for ferromagnetism, and the Lorentz model for motion in a random medium. Schramm–Löwner evolutions (SLE) arise in various contexts. The choice of topics is strongly motivated by modern applications and focuses on areas that merit further research. Special features include a simple account of Smirnov's proof of Cardy's formula for critical percolation, and a fairly full account of the theory of influence and sharp-thresholds.

Accessible to a wide audience of mathematicians and physicists, this book can be used as a graduate course text. Each chapter ends with a range of exercises.

GEOFFREY GRIMMETT is Professor of Mathematical Statistics in the Statistical Laboratory at the University of Cambridge.

INSTITUTE OF MATHEMATICAL STATISTICS
TEXTBOOKS

Probability on Graphs

Random Processes on Graphs and Lattices

GEOFFREY GRIMMETT

Statistical Laboratory
University of Cambridge

CAMBRIDGE
UNIVERSITY PRESS

CAMBRIDGE UNIVERSITY PRESS
Cambridge, New York, Melbourne, Madrid, Cape Town, Singapore,
São Paulo, Delhi, Dubai, Tokyo

Cambridge University Press
The Edinburgh Building, Cambridge CB2 8RU, UK

Published in the United States of America by Cambridge University Press, New York

www.cambridge.org
Information on this title: www.cambridge.org/9780521197984

First published 2010

Printed in the United Kingdom at the University Press, Cambridge

A catalogue record for this publication is available from the British Library

ISBN 978-0-521-19798-4 Hardback
ISBN 978-0-521-14735-4 Paperback

Contents

Preface

Within the menagerie of objects studied in contemporary probability theory, there are a number of related animals that have attracted great interest amongst probabilists and physicists in recent years. The inspiration for many of these objects comes from physics, but the mathematical subject has taken on a life of its own, and many beautiful constructions have emerged. The overall target of these notes is to identify some of these topics, and to develop their basic theory at a level suitable for mathematics graduates.

If the two principal characters in these notes are random walk and percolation, they are only part of the rich theory of uniform spanning trees, self-avoiding walks, random networks, models for ferromagnetism and the spread of disease, and motion in random environments. This is an area that has attracted many fine scientists, by virtue, perhaps, of its special mixture of modelling and problem-solving. There remain many open problems. It is the experience of the author that these may be explained successfully to a graduate audience open to inspiration and provocation.

The material described here may be used for personal study, and as the bases of lecture courses of between 24 and 48 hours duration. Little is assumed about the mathematical background of the audience beyond some basic probability theory, but students should be willing to get their hands dirty if they are to profit. Care should be taken in the setting of examinations, since problems can be unexpectedly difficult. Successful examinations may be designed, and some help is offered through the inclusion of exercises at the ends of chapters. As an alternative to a conventional examination, students may be asked to deliver presentations on aspects and extensions of the topics studied.

Chapter 1 is devoted to the relationship between random walks (on graphs) and electrical networks. This leads to the Thomson and Rayleigh principles, and thence to a proof of Pólya's theorem. In Chapter 2, we describe Wilson's algorithm for constructing a uniform spanning tree (UST), and we discuss boundary conditions and weak limits for UST on a lattice. This chapter includes a brief introduction to Schramm–Löwner evolutions (SLE).

Percolation theory appears first in Chapter 3, together with a short introduction to self-avoiding walks. Correlation inequalities and other general
techniques are described in Chapter 4. A special feature of this part of the
book is a fairly full treatment of influence and sharp-threshold theorems for
product measures, and more generally for monotone measures.

We return to the basic theory of percolation in Chapter 5, including a full
account of Smirnov's proof of Cardy's formula. This is followed in Chapter
6 by a study of the contact model on lattices and trees.

Chapter 7 begins with a proof of the equivalence of Gibbs states and
Markov fields, and continues with an introduction to the Ising and Potts
models. Chapter 8 is an account of the random-cluster model. The quantum
Ising model features in the next chapter, particularly through its relationship
to a continuum random-cluster model, and the consequent analysis using
stochastic geometry.

Interacting particle systems form the basis of Chapter 10. This is a large
field in its own right, and little is done here beyond introductions to the
contact, voter, exclusion models, and the stochastic Ising model. Chapter
11 is devoted to random graphs of Erdős–Rényi type. There are accounts
of the giant cluster, and of the chromatic number via an application of
Hoeffding's inequality for the tail of a martingale.

The final Chapter 12 contains one of the most notorious open problems
in stochastic geometry, namely the Lorentz model (or Ehrenfest wind–tree
model) on the square lattice.

These notes are based in part on courses given by the author within Part
3 of the Mathematical Tripos at Cambridge University over a period of several years. They have been prepared in this form as background material for
lecture courses presented to outstanding audiences of students and profes-
sors at the 2008 PIMS–UBC Summer School in Probability, and during the
programme on Statistical Mechanics at the Institut Henri Poincaré, Paris,
during the last quarter of 2008. They were written in part during a visit
to the Mathematics Department at UCLA (with partial support from NSF
grant DMS-0301795), to which the author expresses his gratitude for the
warm welcome received there, and in part during programmes at the Isaac
Newton Institute and the Institut Henri Poincaré–Centre Emile Borel.

Throughout this work, pointers are included to more extensive accounts
of the topics covered. The selection of references is intended to be useful
rather than comprehensive.

The author thanks four artists for permission to include their work: Tom
Kennedy (Fig. 2.1), Oded Schramm (Figs 2.2–2.4), Raphaël Cerf (Fig. 5.3),
and Julien Dubédat (Fig. 5.18). The section on influence has benefited

from conversations with Rob van den Berg and Tom Liggett. Stanislav Smirnov and Wendelin Werner have consented to the inclusion of some of their neat arguments, hitherto unpublished. Several readers have proposed suggestions and corrections. Thank you, everyone!

G. R. G.
Cambridge
April 2010

1

Random walks on graphs

The theory of electrical networks is a fundamental tool for studying
the recurrence of reversible Markov chains. The Kirchhoff laws and
Thomson principle permit a neat proof of Pólya's theorem for random
walk on a d-dimensional grid.

1.1 Random walks and reversible Markov chains

A basic knowledge of probability theory is assumed in this volume. Readers
keen to acquire this are referred to [122] for an elementary introduction, and
to [121] for a somewhat more advanced account. We shall generally use the
letter $\mathbb{P}$ to denote a generic probability measure, with more specific notation
when helpful. The expectation of a random variable f will be written as
either $\mathbb{P}(f)$ or $\mathbb{E}(f)$.

Only a little knowledge is assumed about graphs, and many readers will
have sufficient acquaintance already. Others are advised to consult Section
1.6. Of the many books on graph theory, we mention [43].

Let $G = (V, E)$ be a finite or countably infinite graph, which we assume
for simplicity to have neither loops nor multiple edges. If G is infinite, we
shall usually assume in addition that every vertex-degree is finite. A particle
moves around the vertex-set V. Having arrived at the vertex S_n at time n, its
next position S_{n+1} is chosen uniformly at random from the set of neighbours
of S_n. The trajectory of the particle is called a *simple random walk* (SRW)
on G.

Two of the basic questions concerning simple random walk are:

1. Under what conditions is the walk *recurrent*, in that it returns (almost
   surely) to its starting point?
2. How does the distance between S_n and S_0 behave as $n \to \infty$?

The above SRW is symmetric in that the jumps are chosen *uniformly*
from the set of available neighbours. In a more general process, we take a
function $w : E \to (0, \infty)$, and we jump along the edge e with probability
proportional to w_e.

Any reversible Markov chain[1] on the set V gives rise to such a walk as follows. Let $Z = (Z_n : n \geq 0)$ be a Markov chain on V with transition matrix P, and assume that Z is reversible with respect to some positive function $\pi : V \to (0, \infty)$, which is to say that

$$(1.1) \qquad \pi_u p_{u,v} = \pi_v p_{v,u}, \qquad u, v \in V.$$

With each distinct pair $u, v \in V$, we associate the weight

$$(1.2) \qquad w_{u,v} = \pi_u p_{u,v},$$

noting by (1.1) that $w_{u,v} = w_{v,u}$. Then

$$(1.3) \qquad p_{u,v} = \frac{w_{u,v}}{W_u}, \qquad u, v \in V,$$

where

$$W_u = \sum_{v \in V} w_{u,v}, \qquad u \in V.$$

That is, given that $Z_n = u$, the chain jumps to a new vertex v with probability proportional to $w_{u,v}$. This may be set in the context of a random walk on the graph with the vertex-set V, and with edge-set containing all $e = \langle u, v \rangle$ such that $p_{u,v} > 0$. With the edge e we associate the weight $w_e = w_{u,v}$.

In this chapter, we develop the relationship between random walks on G and electrical networks on G. There are some excellent accounts of this area, and the reader is referred to the books of Doyle and Snell [72], Lyons and Peres [181], and Aldous and Fill [18], amongst others. The connection between these two topics is made via the so-called 'harmonic functions' of the random walk.

1.4 Definition. Let $U \subseteq V$, and let Z be a Markov chain on V with transition matrix P, that is reversible with respect to the positive function π. The function $f : V \to \mathbb{R}$ is *harmonic* on U (with respect to the transition matrix P) if

$$f(u) = \sum_{v \in V} p_{u,v} f(v), \qquad u \in U,$$

or equivalently, if $f(u) = \mathbb{E}(f(Z_1) \mid Z_0 = u)$ for $u \in U$.

From the pair (P, π), we can construct the graph G as above, and the weight function w as in (1.2). We refer to the pair (G, w) as the weighted graph associated with (P, π). We shall speak of f as being harmonic (for (G, w)) if it is harmonic with respect to P.

---

[1] An account of the basic theory of Markov chains may be found in [121].

The so-called hitting probabilities are the basic examples of harmonic functions for the chain Z. Let $U \subseteq V$, $W = V \setminus U$, and $s \in U$. For $u \in U$, let $g(u)$ be the probability that the chain, started at u, hits s before W. That is,

$$g(u) = \mathbb{P}_u(Z_n = s \text{ for some } n < T_W),$$

where

$$T_W = \inf\{n \geq 0 : Z_n \in W\}$$

is the first-passage time to W, and $\mathbb{P}_u(\cdot) = \mathbb{P}(\cdot \mid Z_0 = u)$ denotes the probability measure conditional on the chain starting at u.

1.5 Theorem. *The function g is harmonic on $U \setminus \{s\}$.*

Evidently, $g(s) = 1$, and $g(v) = 0$ for $v \in W$. We speak of these values of g as being the 'boundary conditions' of the harmonic function g.

Proof. This is an elementary exercise using the Markov property. For $u \notin W \cup \{s\}$,

$$g(u) = \sum_{v \in V} p_{u,v} \mathbb{P}_u\left(Z_n = s \text{ for some } n < T_W \mid Z_1 = v\right)$$

$$= \sum_{v \in V} p_{u,v} g(v),$$

as required. $\qquad\qquad\qquad\qquad\qquad\qquad\qquad\qquad\qquad\qquad\qquad\qquad\square$

1.2 Electrical networks

Throughout this section, $G = (V, E)$ is a finite graph with neither loops nor multiple edges, and $w : E \to (0, \infty)$ is a weight function on the edges. We shall assume further that G is connected.

We may build an electrical network with diagram G, in which the edge e has conductance w_e (or, equivalently, resistance $1/w_e$). Let $s, t \in V$ be distinct vertices termed *sources*, and write $S = \{s, t\}$ for the *source-set*. Suppose we connect a battery across the pair s, t. It is a physical observation that electrons flow along the wires in the network. The flow is described by the so-called Kirchhoff laws, as follows.

To each edge $e = \langle u, v \rangle$, there are associated (directed) quantities $\phi_{u,v}$ and $i_{u,v}$, called the *potential difference* from u to v, and the *current* from u to v, respectively. These are antisymmetric,

$$\phi_{u,v} = -\phi_{v,u}, \qquad i_{u,v} = -i_{v,u}.$$

1.6 Kirchhoff's potential law. The cumulative potential difference around any cycle $v_1, v_2, \ldots, v_n, v_{n+1} = v_1$ of G is zero, that is,

$$(1.7) \qquad \sum_{j=1}^{n} \phi_{v_j, v_{j+1}} = 0.$$

1.8 Kirchhoff's current law. The total current flowing out of any vertex $u \in V$ other than the source-set is zero, that is,

$$(1.9) \qquad \sum_{v \in V} i_{u,v} = 0, \qquad u \neq s, t.$$

The relationship between resistance/conductance, potential difference, and current is given by Ohm's law.

1.10 Ohm's law. For any edge $e = \langle u, v \rangle$,

$$i_{u,v} = w_e \phi_{u,v}.$$

Kirchhoff's potential law is equivalent to the statement that there exists a function $\phi : V \to \mathbb{R}$, called a *potential function*, such that

$$\phi_{u,v} = \phi(v) - \phi(u), \qquad \langle u, v \rangle \in E.$$

Since ϕ is determined up to an additive constant, we are free to pick the potential of any single vertex. Note the convention that *current flows uphill*: $i_{u,v}$ has the same sign as $\phi_{u,v} = \phi(v) - \phi(u)$.

1.11 Theorem. *A potential function is harmonic on the set of vertices other than the source-set.*

Proof. Let $U = V \setminus \{s, t\}$. By Kirchhoff's current law and Ohm's law,

$$\sum_{v \in V} w_{u,v}[\phi(v) - \phi(u)] = 0, \qquad u \in U,$$

which is to say that

$$\phi(u) = \sum_{v \in V} \frac{w_{u,v}}{W_u} \phi(v), \qquad u \in U,$$

where

$$W_u = \sum_{v \in V} w_{u,v}.$$

That is, ϕ is harmonic on U. $\qquad \qquad \square$

We can use Ohm's law to express the potential differences in terms of the currents, and thus the two Kirchhoff laws may be viewed as concerning the currents only. Equation (1.7) becomes

$$(1.12) \qquad \sum_{j=1}^{n} \frac{i_{v_j, v_{j+1}}}{w_{\langle v_j, v_{j+1} \rangle}} = 0,$$

valid for any cycle $v_1, v_2, \ldots, v_n, v_{n+1} = v_1$. With (1.7) written thus, each law is linear in the currents, and the superposition principle follows.

1.13 Theorem (Superposition principle). *If i^1 and i^2 are solutions of the two Kirchhoff laws with the same source-set, then so is the sum $i^1 + i^2$.*

Next we introduce the concept of a 'flow' on the graph.

1.14 Definition. Let $s, t \in V$, $s \neq t$. An *s/t-flow* j is a vector $j = (j_{u,v} : u, v \in V, \ u \neq v)$, such that:
(a) $j_{u,v} = -j_{v,u}$,
(b) $j_{u,v} = 0$ whenever $u \not\sim v$,
(c) for any $u \neq s, t$, we have that $\sum_{v \in V} j_{u,v} = 0$.

The vertices s and t are called the 'source' and 'sink' of an s/t flow, and we usually abbreviate 's/t flow' to 'flow'. For any flow j, we write

$$J_u = \sum_{v \in V} j_{u,v}, \qquad u \in U,$$

noting by (c) above that $J_u = 0$ for $u \neq s, t$. Thus,

$$J_s + J_t = \sum_{u \in V} J_u = \sum_{u,v \in V} j_{u,v} = \tfrac{1}{2} \sum_{u,v \in V} (j_{u,v} + j_{v,u}) = 0.$$

Therefore, $J_s = -J_t$, and we call $|J_s|$ the *size* of the flow j, denoted $|j|$. If $|J_s| = 1$, we call j a *unit flow*. We shall normally take $J_s > 0$, in which case s is the *source*, and t the *sink* of the flow, and we say that j is a flow from s to t.

Note that any solution i to the Kirchhoff laws with source-set $\{s, t\}$ is an s/t flow.

1.15 Theorem. *Let i^1 and i^2 be two solutions of the Kirchhoff laws with the same source and sink and equal size. Then $i^1 = i^2$.*

Proof. By the superposition principle, $j = i^1 - i^2$ satisfies the two Kirchhoff laws. Furthermore, under the flow j, no current enters or leaves the system. Therefore, $J_v = 0$ for all $v \in V$. Suppose $j_{u_1, u_2} > 0$ for some edge $\langle u_1, u_2 \rangle$. By the Kirchhoff current law, there exists u_3 such that

$j_{u_2,u_3} > 0$. By iteration, there exists a cycle $u_1, u_2, \ldots, u_n, u_{n+1} = u_1$ such that $j_{u_j,u_{j+1}} > 0$ for $j = 1, 2, \ldots, n$. By Ohm's law, the corresponding potential function satisfies

$$\phi(u_1) < \phi(u_2) < \cdots < \phi(u_{n+1}) = \phi(u_1),$$

a contradiction. Therefore, $j_{u,v} = 0$ for all u, v. □

For a given size of input current, and given source s and sink t, there can be no more than one solution to the two Kirchhoff laws, but is there a solution at all? The answer is of course affirmative, and the unique solution can be expressed explicitly in terms of counts of spanning trees.[2] Consider first the special case when $w_e = 1$ for all $e \in E$. Let N be the number of spanning trees of G. For any edge $\langle a, b \rangle$, let $\Pi(s, a, b, t)$ be the property of spanning trees that: the unique s/t path in the tree passes along the edge $\langle a, b \rangle$ in the direction from a to b. Let $\mathcal{N}(s, a, b, t)$ be the set of spanning trees of G with the property $\Pi(s, a, b, t)$, and $N(s, a, b, t) = |\mathcal{N}(s, a, b, t)|$.

1.16 Theorem. *The function*

$$(1.17) \qquad i_{a,b} = \frac{1}{N}\left[N(s, a, b, t) - N(s, b, a, t)\right], \qquad \langle a, b \rangle \in E,$$

defines a unit flow from s to t satisfying the Kirchhoff laws.

Let T be a spanning tree of G chosen uniformly at random from the set $\mathcal{T}$ of all such spanning trees. By Theorem 1.16 and the previous discussion, the unique solution to the Kirchhoff laws with source s, sink t, and size 1 is given by

$$i_{a,b} = \mathbb{P}\big(T \text{ has } \Pi(s, a, b, t)\big) - \mathbb{P}\big(T \text{ has } \Pi(s, b, a, t)\big).$$

We shall return to uniform spanning trees in Chapter 2.

We prove Theorem 1.16 next. Exactly the same proof is valid in the case of general conductances w_e. In that case, we define the weight of a spanning tree T as

$$w(T) = \prod_{e \in T} w_e,$$

and we set

$$(1.18) \qquad N^* = \sum_{T \in \mathcal{T}} w(T), \qquad N^*(s, a, b, t) = \sum_{T \text{ with } \Pi(s,a,b,t)} w(T).$$

The conclusion of Theorem 1.16 holds in this setting with

$$i_{a,b} = \frac{1}{N^*}\left[N^*(s, a, b, t) - N^*(s, b, a, t)\right], \qquad \langle a, b \rangle \in E.$$

---

[2]This was discovered in an equivalent form by Kirchhoff in 1847, [153].

Proof of Theorem 1.16. We first check the Kirchhoff current law. In every spanning tree T, there exists a unique vertex b such that the s/t path of T contains the edge $\langle s, b \rangle$, and the path traverses this edge from s to b. Therefore,

$$\sum_{b \in V} N(s, s, b, t) = N, \qquad N(s, b, s, t) = 0 \text{ for } b \in V.$$

By (1.17),

$$\sum_{b \in V} i_{s,b} = 1,$$

and, by a similar argument, $\sum_{b \in V} i_{b,t} = 1$.

Let T be a spanning tree of G. The contribution towards the quantity $i_{a,b}$, made by T, depends on the s/t path π of T, and equals

$$
(1.19) \qquad
\begin{array}{ll}
N^{-1} & \text{if } \pi \text{ passes along } \langle a, b \rangle \text{ from } a \text{ to } b, \\
-N^{-1} & \text{if } \pi \text{ passes along } \langle a, b \rangle \text{ from } b \text{ to } a, \\
0 & \text{if } \pi \text{ does not contain the edge } \langle a, b \rangle.
\end{array}
$$

Let $v \in V$, $v \neq s, t$, and write $I_v = \sum_{w \in V} i_{v,w}$. If $v \in \pi$, the contribution of T towards I_v is $N^{-1} - N^{-1} = 0$, since π arrives at v along some edge of the form $\langle a, v \rangle$, and departs v along some edge of the form $\langle v, b \rangle$. If $v \notin \pi$, then T contributes 0 to I_v. Summing over T, we obtain that $I_v = 0$ for all $v \neq s, t$, as required for the Kirchhoff current law.

We next check the Kirchhoff potential law. Let $v_1, v_2, \ldots, v_n, v_{n+1} = v_1$ be a cycle C of G. We shall show that

$$(1.20) \qquad \sum_{j=1}^{n} i_{v_j, v_{j+1}} = 0,$$

and this will confirm (1.12), on recalling that $w_e = 1$ for all $e \in E$. It is more convenient in this context to work with 'bushes' than spanning trees. A *bush* (or, more precisely, an s/t-*bush*) is defined to be a forest on V containing exactly two trees, one denoted T_s and containing s, and the other denoted T_t and containing t. We write (T_s, T_t) for this bush. Let $e = \langle a, b \rangle$, and let $\mathcal{B}(s, a, b, t)$ be the set of bushes with $a \in T_s$ and $b \in T_t$. The sets $\mathcal{B}(s, a, b, t)$ and $\mathcal{N}(s, a, b, t)$ are in one–one correspondence, since the addition of e to $B \in \mathcal{B}(s, a, b, t)$ creates a unique member $T = T(B)$ of $\mathcal{N}(s, a, b, t)$, and vice versa.

By (1.19) and the above, a bush $B = (T_s, T_t)$ makes a contribution to $i_{a,b}$ of

$$N^{-1} \quad \text{if } B \in \mathcal{B}(s, a, b, t),$$
$$-N^{-1} \quad \text{if } B \in \mathcal{B}(s, b, a, t),$$
$$0 \quad \text{otherwise.}$$

Therefore, B makes a contribution towards the sum in (1.20) that is equal to $N^{-1}(F_+ - F_-)$, where F_+ (respectively, F_-) is the number of pairs v_j, v_{j+1} of C, $1 \le j \le n$, with $v_j \in T_s$, $v_{j+1} \in T_t$ (respectively, $v_{j+1} \in T_s$, $v_j \in T_t$). Since C is a cycle, $F_+ = F_-$, whence each bush contributes 0 to the sum, and (1.20) is proved. □

1.3 Flows and energy

Let $G = (V, E)$ be a connected graph as before. Let $s, t \in V$ be distinct vertices, and let j be an s/t flow. With w_e the conductance of the edge e, the (dissipated) *energy* of j is defined as

$$E(j) = \sum_{e = \langle u, v \rangle \in E} j_{u,v}^2 / w_e = \tfrac{1}{2} \sum_{\substack{u, v \in V \\ u \sim v}} j_{u,v}^2 / w_{\langle u, v \rangle}.$$

The following piece of linear algebra will be useful.

1.21 Proposition. *Let $\psi : V \to \mathbb{R}$, and let j be an s/t flow. Then*

$$[\psi(t) - \psi(s)]J_s = \tfrac{1}{2} \sum_{u, v \in V} [\psi(v) - \psi(u)]j_{u,v}.$$

Proof. By the properties of a flow,

$$\sum_{u, v \in V} [\psi(v) - \psi(u)]j_{u,v} = \sum_{v \in V} \psi(v)(-J_v) - \sum_{u \in V} \psi(u)J_u$$
$$= -2[\psi(s)J_s + \psi(t)J_t]$$
$$= 2[\psi(t) - \psi(s)]J_s,$$

as required. □

Let ϕ and i satisfy the Kirchhoff laws. We apply Proposition 1.21 with $\psi = \phi$ and $j = i$ to find by Ohm's law that

$$(1.22) \qquad\qquad E(i) = [\phi(t) - \phi(s)]I_s.$$

That is, the energy of the true current-flow i between s to t equals the energy dissipated in a single $\langle s, t \rangle$ edge carrying the same potential difference and total current. The conductance W_{eff} of such an edge would satisfy Ohm's law, that is,

$$(1.23) \qquad I_s = W_{\text{eff}}[\phi(t) - \phi(s)],$$

and we define the *effective conductance* W_{eff} by this equation. The effective resistance is

$$(1.24) \qquad R_{\text{eff}} = \frac{1}{W_{\text{eff}}},$$

which, by (1.22)–(1.23), equals $E(i)/I_s^2$. We state this as a lemma.

1.25 Lemma. *The effective resistance R_{eff} of the network between vertices s and t equals the dissipated energy when a unit flow passes from s to t.*

It is useful to be able to do calculations. Electrical engineers have devised a variety of formulaic methods for calculating the effective resistance of a network, of which the simplest are the series and parallel laws, illustrated in Figure 1.1.

Figure 1.1. Two edges e and f in parallel and in series.

1.26 Series law. Two resistors of size r_1 and r_2 in series may be replaced by a single resistor of size $r_1 + r_2$.

1.27 Parallel law. Two resistors of size r_1 and r_2 in parallel may be replaced by a single resistor of size R where $R^{-1} = r_1^{-1} + r_2^{-1}$.

A third such rule, the so-called 'star–triangle transformation', may be found at Exercise 1.5. The following 'variational principle' has many uses.

1.28 Theorem (Thomson principle). *Let $G = (V, E)$ be a connected graph, and w_e, $e \in E$, (strictly positive) conductances. Let $s, t \in V$, $s \neq t$. Amongst all unit flows through G from s to t, the flow that satisfies the Kirchhoff laws is the unique s/t flow i that minimizes the dissipated energy. That is,*

$$E(i) = \inf\{E(j) : j \text{ a unit flow from } s \text{ to } t\}.$$

Proof. Let j be a unit flow from source s to sink t, and set $k = j - i$ where i is the (unique) unit-flow solution to the Kirchhoff laws. Thus, k is a flow with zero size. Now, with $e = \langle u, v \rangle$ and $r_e = 1/w_e$,

$$2E(j) = \sum_{u,v \in V} j_{u,v}^2 r_e = \sum_{u,v \in V} (k_{u,v} + i_{u,v})^2 r_e$$

$$= \sum_{u,v \in V} k_{u,v}^2 r_e + \sum_{u,v \in V} i_{u,v}^2 r_e + 2 \sum_{u,v \in V} i_{u,v} k_{u,v} r_e.$$

Let ϕ be the potential function corresponding to i. By Ohm's law and Proposition 1.21,

$$\sum_{u,v \in V} i_{u,v} k_{u,v} r_e = \sum_{u,v \in V} [\phi(v) - \phi(u)] k_{u,v}$$

$$= 2[\phi(t) - \phi(s)] K_s,$$

which equals zero. Therefore, $E(j) \geq E(i)$, with equality if and only if $j = i$. $\qquad\square$

The Thomson 'variational principle' leads to a proof of the 'obvious' fact that the effective resistance of a network is a non-decreasing function of the resistances of individual edges.

1.29 Theorem (Rayleigh principle). *The effective resistance R_{eff} of the network is a non-decreasing function of the edge-resistances $(r_e : e \in E)$.*

It is left as an exercise to show that R_{eff} is a concave function of the (r_e). See Exercise 1.6.

Proof. Consider two vectors $(r_e : e \in E)$ and $(r'_e : e \in E)$ of edge-resistances with $r_e \leq r'_e$ for all e. Let i and i' denote the corresponding unit flows satisfying the Kirchhoff laws. By Lemma 1.25, with $r_e = r_{\langle u,v \rangle}$,

$$R_{\text{eff}} = \frac{1}{2} \sum_{\substack{u,v \in V \\ u \sim v}} i_{u,v}^2 r_e$$

$$\leq \frac{1}{2} \sum_{\substack{u,v \in V \\ u \sim v}} (i'_{u,v})^2 r_e \qquad \text{by the Thomson principle}$$

$$\leq \frac{1}{2} \sum_{\substack{u,v \in V \\ u \sim v}} (i'_{u,v})^2 r'_e \qquad \text{since } r_e \leq r'_e$$

$$= R'_{\text{eff}},$$

as required. $\qquad\square$

1.4 Recurrence and resistance

Let $G = (V, E)$ be an infinite connected graph with finite vertex-degrees, and let $(w_e : e \in E)$ be (strictly positive) conductances. We shall consider a reversible Markov chain $Z = (Z_n : n \geq 0)$ on the state space V with transition probabilities given by (1.3). Our purpose is to establish a condition on the pair (G, w) that is equivalent to the recurrence of Z.

Let 0 be a distinguished vertex of G, called the 'origin', and suppose $Z_0 = 0$. The graph-theoretic distance between two vertices u, v is the number of edges in a shortest path between u and v, denoted $\delta(u, v)$. Let

$$\Lambda_n = \{u \in V : \delta(0, v) \leq n\},$$
$$\partial \Lambda_n = \Lambda_n \setminus \Lambda_{n-1} = \{u \in V : \delta(0, v) = n\}.$$

We think of $\partial \Lambda_n$ as the 'boundary' of Λ_n. Let G_n be the subgraph of G comprising the vertex-set Λ_n, together with all edges between them. We let $\overline{G}_n$ be the graph obtained from G_n by identifying all vertices in $\partial \Lambda_n$, and we denote the identified vertex as I_n. The resulting finite graph $\overline{G}_n$ may be considered an electrical network with sources 0 and I_n. Let $R_{\text{eff}}(n)$ be the effective resistance of this network. The graph $\overline{G}_n$ may be obtained from $\overline{G}_{n+1}$ by identifying all vertices lying in $\partial \Lambda_n \cup \{I_{n+1}\}$, and thus, by the Rayleigh principle, $R_{\text{eff}}(n)$ is non-decreasing in n. Therefore, the limit

$$R_{\text{eff}} = \lim_{n \to \infty} R_{\text{eff}}(n)$$

exists.

1.30 Theorem. *The probability of ultimate return by Z to the origin 0 is given by*

$$\mathbb{P}_0(Z_n = 0 \text{ for some } n \geq 1) = 1 - \frac{1}{W_0 R_{\text{eff}}},$$

where $W_0 = \sum_{v: v \sim 0} w_{\langle 0, v \rangle}$.

The return probability is non-decreasing if $W_0 R_{\text{eff}}$ is increased. By the Rayleigh principle, this can be achieved, for example, by removing an edge of E that is not incident to 0. The removal of an edge incident to 0 can have the opposite effect, since W_0 decreases while R_{eff} increases (see Figure 1.2).

A $0/\infty$ *flow* is a vector $j = (j_{u,v} : u, v \in V, u \neq v)$ satisfying (1.14)(a)–(b) and also (c) for all $u \neq 0$. That is, it has source 0 but no sink.

1.31 Corollary.

(a) *The chain Z is recurrent if and only if $R_{\text{eff}} = \infty$.*

(b) *The chain Z is transient if and only if there exists a non-zero $0/\infty$ flow j on G whose energy $E(j) = \sum_e j_e^2 / w_e$ satisfies $E(j) < \infty$.*

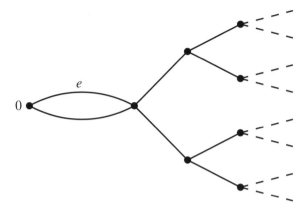

Figure 1.2. This is an infinite binary tree with two parallel edges joining the origin to the root. When each edge has unit resistance, it is an easy calculation that $R_{\text{eff}} = \frac{3}{2}$, so the probability of return to 0 is $\frac{2}{3}$. If the edge e is removed, this probability becomes $\frac{1}{2}$.

It is left as an exercise to extend this to countable graphs G without the assumption of finite vertex-degrees.

Proof of Theorem 1.30. Let

$$g_n(v) = \mathbb{P}_v(Z \text{ hits } \partial \Lambda_n \text{ before } 0), \qquad v \in \Lambda_n.$$

By Theorem 1.5, g_n is the unique harmonic function on G_n with boundary conditions

$$g_n(0) = 0, \qquad g_n(v) = 1 \text{ for } v \in \partial \Lambda_n.$$

Therefore, g_n is a potential function on $\overline{G}_n$ viewed as an electrical network with source 0 and sink I_n.

By conditioning on the first step of the walk, and using Ohm's law,

$$\mathbb{P}_0(Z \text{ returns to } 0 \text{ before reaching } \partial \Lambda_n)$$

$$= 1 - \sum_{v:\, v \sim 0} p_{0,v} g_n(v)$$

$$= 1 - \sum_{v:\, v \sim 0} \frac{w_{0,v}}{W_0} [g_n(v) - g_n(0)]$$

$$= 1 - \frac{|i(n)|}{W_0},$$

where $i(n)$ is the flow of currents in $\overline{G}_n$, and $|i(n)|$ is its size. By (1.23)–(1.24), $|i(n)| = 1/R_{\text{eff}}(n)$. The theorem is proved on noting that

$$\mathbb{P}_0(Z \text{ returns to } 0 \text{ before reaching } \partial \Lambda_n) \to \mathbb{P}_0(Z_n = 0 \text{ for some } n \geq 1)$$

as $n \to \infty$, by the continuity of probability measures.   □

Proof of Corollary 1.31.   Part (a) is an immediate consequence of Theorem 1.30, and we turn to part (b). By Lemma 1.25, there exists a unit flow $i(n)$ in $\overline{G}_n$ with source 0 and sink I_n, and with energy $E(i(n)) = R_{\text{eff}}(n)$. Let i be a non-zero $0/\infty$ flow; by dividing by its size, we may take i to be a unit flow. When restricted to the edge-set E_n of $\overline{G}_n$, i forms a unit flow from 0 to I_n. By the Thomson principle, Theorem 1.28,

$$E(i(n)) \leq \sum_{e \in E_n} i_e^2 / w_e \leq E(i),$$

whence

$$E(i) \geq \lim_{n \to \infty} E(i(n)) = R_{\text{eff}}.$$

Therefore, by part (a), $E(i) = \infty$ if the chain is recurrent.

Suppose, conversely, that the chain is transient. By diagonal selection[3], there exists a subsequence (n_k) along which $i(n_k)$ converges to some limit j (that is, $i(n_k)_e \to j_e$ for every $e \in E$). Since each $i(n_k)$ is a unit flow from the origin, j is a unit $0/\infty$ flow. Now,

$$
\begin{aligned}
E(i(n_k)) &= \sum_{e \in E} i(n_k)_e^2 / w_e \\
&\geq \sum_{e \in E_m} i(n_k)_e^2 / w_e \\
&\to \sum_{e \in E_m} j(e)^2 / w_e \qquad \text{as } k \to \infty \\
&\to E(j) \qquad\qquad\quad \text{as } m \to \infty.
\end{aligned}
$$

Therefore,

$$E(j) \leq \lim_{k \to \infty} R_{\text{eff}}(n_k) = R_{\text{eff}} < \infty,$$

and j is a flow with the required properties.   □

---

[3] *Diagonal selection:* Let $(x_m(n) : m, n \geq 1)$ be a bounded collection of reals. There exists an increasing sequence $n_1, n_2, \dots$ of positive integers such that, for every m, the limit $\lim_{k \to \infty} x_m(n_k)$ exists.

1.5 Pólya's theorem

The d-dimensional cubic lattice $\mathbb{L}^d$ has vertex-set $\mathbb{Z}^d$ and edges between any two vertices that are Euclidean distance one apart. The following celebrated theorem can be proved by estimating effective resistances.[4]

1.32 Pólya's Theorem [200]. *Symmetric random walk on the lattice $\mathbb{L}^d$ in d dimensions is recurrent if $d = 1, 2$ and transient if $d \geq 3$.*

The advantage of the following proof of Pólya's theorem over more standard arguments is its robustness with respect to the underlying graph. Similar arguments are valid for graphs that are, in broad terms, comparable to $\mathbb{L}^d$ when viewed as electrical networks.

Proof. For simplicity, and with only little loss of generality (see Exercise 1.10), we shall concentrate on the cases $d = 2, 3$. Let $d = 2$, for which case we aim to show that $R_{\mathrm{eff}} = \infty$. This is achieved by finding an infinite lower bound for R_{eff}, and lower bounds can be obtained by decreasing individual edge-resistances. The identification of two vertices of a network amounts to the addition of a resistor with 0 resistance, and, by the Rayleigh principle, the effective resistance of the network can only decrease.

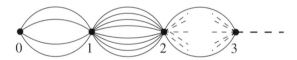

Figure 1.3.   The vertex labelled i is a composite vertex obtained by identifying all vertices with distance i from 0. There are $8i - 4$ edges of $\mathbb{L}^2$ joining vertices $i - 1$ and i.

From $\mathbb{L}^2$, we construct a new graph in which, for each $k = 1, 2, \ldots$, the set $\partial \Lambda_k = \{v \in \mathbb{Z}^2 : \delta(0, v) = k\}$ is identified as a singleton. This transforms $\mathbb{L}^2$ into the graph shown in Figure 1.3. By the series/parallel laws and the Rayleigh principle,

$$R_{\mathrm{eff}}(n) \geq \sum_{i=1}^{n-1} \frac{1}{8i - 4},$$

whence $R_{\mathrm{eff}}(n) \geq c \log n \to \infty$ as $n \to \infty$.

Suppose now that $d = 3$. There are at least two ways of proceeding. We shall present one such route taken from [182], and we shall then sketch

---

[4]An amusing story is told in [201] about Pólya's inspiration for this theorem.

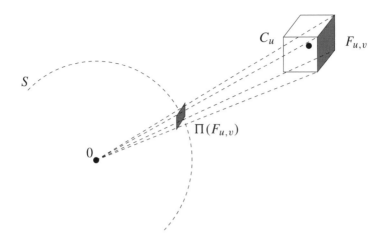

Figure 1.4.   The flow along the edge $\langle u, v \rangle$ is equal to the area of the projection $\Pi(F_{u,v})$ on the unit sphere centred at the origin, with a suitable convention for its sign.

the second which has its inspiration in [72]. By Corollary 1.31, it suffices to construct a non-zero $0/\infty$ flow with finite energy. Let S be the surface of the unit sphere of $\mathbb{R}^3$ with centre at the origin 0. Take $u \in \mathbb{Z}^3$, $u \neq 0$, and position a unit cube C_u in $\mathbb{R}^3$ with centre at u and edges parallel to the axes (see Figure 1.4). For each neighbour v of u, the directed edge $[u, v\rangle$ intersects a unique face, denoted $F_{u,v}$, of C_u.

For $x \in \mathbb{R}^3$, $x \neq 0$, let $\Pi(x)$ be the point of intersection with S of the straight line segment from 0 to x. Let $j_{u,v}$ be equal in absolute value to the surface measure of $\Pi(F_{u,v})$. The sign of $j_{u,v}$ is taken to be positive if and only if the scalar product of $\frac{1}{2}(u + v)$ and $v - u$, viewed as vectors in $\mathbb{R}^3$, is positive. Let $j_{v,u} = -j_{u,v}$. We claim that j is a $0/\infty$ flow on $\mathbb{L}^3$. Parts (a) and (b) of Definition 1.14 follow by construction, and it remains to check (c).

The surface of C_u has a projection $\Pi(C_u)$ on S. The sum $J_u = \sum_{v \sim u} j_{u,v}$ is the integral over $\mathbf{x} \in \Pi(C_u)$, with respect to surface measure, of the number of neighbours v of u (counted with sign) for which $\mathbf{x} \in \Pi(F_{u,v})$. Almost every $\mathbf{x} \in \Pi(C_u)$ is counted twice, with signs $+$ and $-$. Thus the integral equals 0, whence $J_u = 0$ for all $u \neq 0$.

It is easily seen that $J_0 \neq 0$, so j is a non-zero flow. Next, we estimate its energy. By an elementary geometric consideration, there exist $c_i < \infty$ such that:

(i) $|j_{u,v}| \leq c_1/|u|^2$ for $u \neq 0$, where $|u| = \delta(0, u)$ is the length of a shortest path from 0 to u,

(ii) the number of $u \in \mathbb{Z}^3$ with $|u| = n$ is smaller than $c_2 n^2$.

It follows that

$$E(j) \leq \sum_{u \neq 0} \sum_{v \sim u} j_{u,v}^2 \leq \sum_{n=1}^{\infty} 6c_2 n^2 \left(\frac{c_1}{n^2}\right)^2 < \infty,$$

as required. ☐

Another way of showing $R_{\text{eff}} < \infty$ when $d = 3$ is to find a finite upper bound for R_{eff}. Upper bounds can be obtained by increasing individual edge-resistances, or by removing edges. The idea is to embed a tree with finite resistance in $\mathbb{L}^3$. Consider a binary tree T_ρ in which each connection between generation $n-1$ and generation n has resistance ρ^n, where $\rho > 0$. It is an easy exercise using the series/parallel laws that the effective resistance between the root and infinity is

$$R_{\text{eff}}(T_\rho) = \sum_{n=1}^{\infty} (\rho/2)^n,$$

which we make finite by choosing $\rho < 2$. We proceed to embed T_ρ in $\mathbb{Z}^3$ in such a way that a connection between generation $n-1$ and generation n is a lattice-path of length order ρ^n. There are 2^n vertices of T_ρ in generation n, and their lattice-distance from 0 has order $\sum_{k=1}^{n} \rho^k$, that is, order ρ^n. The surface of the k-ball in $\mathbb{R}^3$ has order k^2, and thus it is necessary that

$$c(\rho^n)^2 \geq 2^n,$$

which is to say that $\rho > \sqrt{2}$.

Let $\sqrt{2} < \rho < 2$. It is now fairly simple to check that $R_{\text{eff}} < c' R_{\text{eff}}(T_\rho)$. This method has been used in [114] to prove the transience of the infinite open cluster of percolation on $\mathbb{L}^3$. It is related to, but different from, the tree embeddings of [72].

1.6 Graph theory

A graph $G = (V, E)$ comprises a finite or countably infinite vertex-set V and an associated edge-set E. Each element of E is an unordered pair u, v of vertices written $\langle u, v \rangle$. Two edges with the same vertex-pairs are said to be in *parallel*, and edges of the form $\langle u, u \rangle$ are called *loops*. The graphs of these notes will generally contain neither parallel edges nor loops, and this is assumed henceforth. Two vertices u, v are said to be joined (or connected) by an edge if $\langle u, v \rangle \in E$. In this case, u and v are the *endvertices* of e, and we write $u \sim v$ and say that u is *adjacent* to v. An edge e is said to be *incident* to its endvertices. The number of edges incident to vertex u is called the *degree* of u, denoted $\deg(u)$. The negation of the relation $\sim$ is written $\nsim$.

Since the edges are unordered pairs, we call such a graph *undirected* (or *unoriented*). If some or all of its edges are *ordered* pairs, written $[u, v\rangle$, the graph is called *directed* (or *oriented*).

A *path* of G is defined as an alternating sequence $v_0, e_0, v_1, e_1, \ldots, e_{n-1}$, v_n of distinct vertices v_i and edges $e_i = \langle v_i, v_{i+1} \rangle$. Such a path has *length* n; it is said to connect v_0 to v_n, and is called a v_0/v_n path. A *cycle* or *circuit* of G is an alternating sequence $v_0, e_0, v_1, \ldots, e_{n-1}, v_n, e_n, x_0$ of vertices and edges such that $v_0, e_0, \ldots, e_{n-1}, v_n$ is a path and $e_n = \langle v_n, v_0 \rangle$. Such a cycle has length $n + 1$. The (graph-theoretic) distance $\delta(u, v)$ from u to v is defined to be the number of edges in a shortest path of G from u to v.

We write $u \longleftrightarrow v$ if there exists a path connecting u and v. The relation $\longleftrightarrow$ is an equivalence relation, and its equivalence classes are called *components* (or *clusters*) of G. The components of G may be considered as either sets of vertices, or graphs. The graph G is *connected* if it has a unique component. It is a *forest* if it contains no cycle, and a *tree* if in addition it is connected.

A *subgraph* of the graph $G = (V, E)$ is a graph $H = (W, F)$ with $W \subseteq V$ and $F \subseteq E$. The subgraph H is a *spanning tree* of G if $V = W$ and H is a tree. A subset $U \subseteq V$ of the vertex-set of G has *boundary* $\partial U = \{u \in U : u \sim v \text{ for some } v \in V \setminus U\}$.

The lattice-graphs are the most important for applications in areas such as statistical mechanics. Lattices are sometimes termed 'crystalline' since they are periodic structures of crystal-like units. A general definition of a lattice may confuse readers more than help them, and instead we describe some principal examples.

Let d be a positive integer. We write $\mathbb{Z} = \{\ldots, -1, 0, 1, \ldots\}$ for the set of all integers, and $\mathbb{Z}^d$ for the set of all d-vectors $v = (v_1, v_2, \ldots, v_d)$ with integral coordinates. For $v \in \mathbb{Z}^d$, we generally write v_i for the ith coordinate of v, and we define

$$\delta(u, v) = \sum_{i=1}^{d} |u_i - v_i|.$$

The *origin* of $\mathbb{Z}^d$ is denoted by 0. We turn $\mathbb{Z}^d$ into a graph, called the *d-dimensional (hyper)cubic lattice*, by adding edges between all pairs u, v of points of $\mathbb{Z}^d$ with $\delta(u, v) = 1$. This graph is denoted as $\mathbb{L}^d$, and its edge-set as $\mathbb{E}^d$: thus, $\mathbb{L}^d = (\mathbb{Z}^d, \mathbb{E}^d)$. We often think of $\mathbb{L}^d$ as a graph embedded

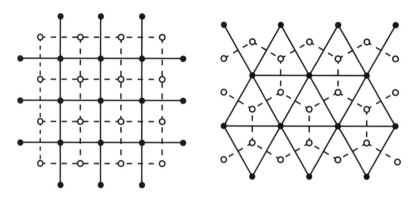

Figure 1.5. The square, triangular, and hexagonal (or 'honeycomb') lattices. The solid and dashed lines illustrate the concept of 'planar duality' discussed on page 41.

in $\mathbb{R}^d$, the edges being straight line-segments between their endvertices. The *edge-set* $\mathbb{E}_V$ of $V \subseteq \mathbb{Z}^d$ is the set of all edges of $\mathbb{L}^d$ both of whose endvertices lie in V.

The two-dimensional cubic lattice $\mathbb{L}^2$ is called the *square lattice* and is illustrated in Figure 1.5. Two other lattices in two dimensions that feature in these notes are drawn there also.

1.7 Exercises

1.1 Let $G = (V, E)$ be a finite connected graph with unit edge-weights. Show that the effective resistance between two distinct vertices s, t of the associated electrical network may be expressed as B/N, where B is the number of s/t-bushes of G, and N is the number of its spanning trees. (See the proof of Theorem 1.16 for an explanation of the term 'bush'.)

Extend this result to general positive edge-weights w_e.

1.2 Let $G = (V, E)$ be a finite connected graph with positive edge-weights $(w_e : e \in E)$, and let N^* be given by (1.18). Show that

$$i_{a,b} = \frac{1}{N^*}\left[N^*(s, a, b, t) - N^*(s, b, a, t) \right]$$

constitutes a unit flow through G from s to t satisfying Kirchhoff's laws.

1.3 (continuation) Let $G = (V, E)$ be finite and connected with given conductances $(w_e : e \in E)$, and let $(x_v : v \in V)$ be reals satisfying $\sum_v x_v = 0$. To G we append a notional vertex labelled ∞, and we join ∞ to each $v \in V$. Show that there exists a solution i to Kirchhoff's laws on the expanded graph, viewed as two laws concerning current flow, such that the current along the edge $\langle v, \infty \rangle$ is x_v.

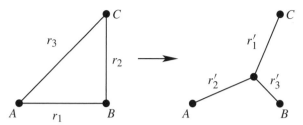

Figure 1.6. Edge-resistances in the star–triangle transformation. The triangle T on the left is replaced by the star S on the right, and the corresponding resistances are as marked.

1.4 Prove the series and parallel laws for electrical networks.

1.5 *Star–triangle transformation.* The triangle T is replaced by the star S in an electrical network, as illustrated in Figure 1.6. Explain the sense in which the two networks are the same, when the resistances are chosen such that $r_j r'_j = c$ for $j = 1, 2, 3$ and some $c = c(r_1, r_2, r_3)$ to be determined.

1.6 Let $R(r)$ be the effective resistance between two given vertices of a finite network with edge-resistances $r = (r(e) : e \in E)$. Show that R is concave in that

$$\tfrac{1}{2}\big[R(r_1) + R(r_2)\big] \le R\big(\tfrac{1}{2}(r_1 + r_2)\big).$$

1.7 *Maximum principle.* Let $G = (V, E)$ be a finite or infinite network with finite vertex-degrees and associated conductances $(w_e : e \in E)$. Let $H = (W, F)$ be a connected subgraph of G, and write

$$\Delta W = \{v \in V \setminus W : v \sim w \text{ for some } w \in W\}$$

for the 'external boundary' of W. Let $\phi : V \to [0, \infty)$ be harmonic on the set W, and suppose the supremum of ϕ on W is achieved and satisfies

$$\sup_{w \in W} \phi(w) = \|\phi\|_\infty := \sup_{v \in V} \phi(v).$$

Show that ϕ is constant on $W \cup \Delta W$, where it takes the value $\|\phi\|_\infty$.

1.8 Let G be an infinite connected graph, and let $\partial \Lambda_n$ be the set of vertices distance n from the vertex labelled 0. With E_n the number of edges joining $\partial \Lambda_n$ to $\partial \Lambda_{n+1}$, show that random walk on G is recurrent if $\sum_n E_n^{-1} = \infty$.

1.9 (continuation) Assume that G is 'spherically symmetric' in that: for all n, for all $x, y \in \partial \Lambda_n$, there exists a graph automorphism that fixes 0 and maps x to y. Show that random walk on G is transient if $\sum_n E_n^{-1} < \infty$.

1.10 Let G be a countably infinite connected graph with finite vertex-degrees, and with a nominated vertex 0. Let H be a connected subgraph of G containing 0. Show that simple random walk, starting at 0, is recurrent on H whenever it is recurrent on G, but that the converse need not hold.

1.11 Let G be a finite connected network with positive conductances $(w_e : e \in E)$, and let a, b be distinct vertices. Let i_{xy} denote the current along an edge from x to y when a unit current flows from the source vertex a to the sink vertex b. Run the associated Markov chain, starting at a, until it reaches b for the first time, and let $u_{x,y}$ be the mean of the total number of transitions of the chain between x and y. Transitions from x to y count positive, and from y to x negative, so that $u_{x,y}$ is the mean number of transitions from x to y, minus the mean number from y to x. Show that $i_{x,y} = u_{x,y}$.

1.12 [72] Let G be an infinite connected graph with bounded vertex-degrees. Let $k \geq 1$, and let G_k be obtained from G by adding an edge between any pair of vertices that are non-adjacent (in G) but separated by graph-theoretic distance k or less. (The graph G_k is sometimes called the *k-fuzz* of G.) Show that simple random walk is recurrent on G_k if and only if it is recurrent on G.

2

Uniform spanning tree

The Uniform Spanning Tree (UST) measure has a property of negative association. A similar property is conjectured for Uniform Forest and Uniform Connected Subgraph. Wilson's algorithm uses loop-erased random walk (LERW) to construct a UST. The UST on the d-dimensional cubic lattice may be defined as the weak limit of the finite-volume measures. When $d = 2$, the corresponding LERW (respectively, UST) converges in a certain manner to the Schramm–Löwner evolution process SLE_2 (respectively, SLE_8) as the grid size approaches zero.

2.1 Definition

Let $G = (V, E)$ be a finite connected graph, and write $\mathcal{T}$ for the set of all spanning trees of G. Let T be picked uniformly at random from $\mathcal{T}$. We call T a *uniform spanning tree*, abbreviated to UST. It is governed by the uniform measure

$$\mathbb{P}(T = t) = \frac{1}{|\mathcal{T}|}, \qquad t \in \mathcal{T}.$$

We may think of T either as a random graph, or as a random subset of E. In the latter case, T may be thought of as a random element of the set $\Omega = \{0, 1\}^E$ of $0/1$ vectors indexed by E.

It is fundamental that UST has a property of *negative association*. In its simplest form, this property may be expressed as follows.

2.1 Theorem. *For $f, g \in E$, $f \neq g$,*

$$(2.2) \qquad \mathbb{P}(f \in T \mid g \in T) \leq \mathbb{P}(f \in T).$$

The proof makes striking use of the Thomson principle via the monotonicity of effective resistance. We obtain the following by a mild extension of the proof. For $B \subseteq E$ and $g \in E \setminus B$,

$$(2.3) \qquad \mathbb{P}(B \subseteq T \mid g \in T) \leq \mathbb{P}(B \subseteq T).$$

21

Proof. Consider G as an electrical network in which each edge has resistance 1. Denote by $i = (i_{v,w} : v, w \in V)$ the current flow in G when a unit current enters at x and leaves at y, and let ϕ be the corresponding potential function. Let $e = \langle x, y \rangle$. By Theorem 1.16,

$$i_{x,y} = \frac{N(x, x, y, y)}{N},$$

where $N(x, x, y, y)$ is the number of spanning trees of G with the property that the unique x/y path passes along the edge e in the direction from x to y, and $N = |\mathcal{T}|$. Therefore, $i_{x,y} = \mathbb{P}(e \in T)$. Since $\langle x, y \rangle$ has unit resistance, $i_{x,y}$ equals the potential difference $\phi(y) - \phi(x)$. By (1.22),

(2.4) $$\mathbb{P}(e \in T) = R^G_{\text{eff}}(x, y),$$

the effective resistance of G between x and y.

Let f, g be distinct edges, and write $G.g$ for the graph obtained from G by contracting g to a single vertex. Contraction provides a one–one correspondence between spanning trees of G containing g, and spanning trees of $G.g$. Therefore, $\mathbb{P}(f \in T \mid g \in T)$ is simply the proportion of spanning trees of $G.g$ containing f. By (2.4),

$$\mathbb{P}(f \in T \mid g \in T) = R^{G.g}_{\text{eff}}(x, y).$$

By the Rayleigh principle, Theorem 1.29,

$$R^{G.g}_{\text{eff}}(x, y) \le R^G_{\text{eff}}(x, y),$$

and the theorem is proved.   □

Theorem 2.1 has been extended by Feder and Mihail [85] to more general 'increasing' events. Let $\Omega = \{0, 1\}^E$, the set of 0/1 vectors indexed by E, and denote by $\omega = (\omega(e) : e \in E)$ a typical member of Ω. The partial order $\le$ on Ω is the usual pointwise ordering: $\omega \le \omega'$ if $\omega(e) \le \omega'(e)$ for all $e \in E$. A subset $A \subseteq \Omega$ is called *increasing* if: for all $\omega, \omega' \in \Omega$ satisfying $\omega \le \omega'$, we have that $\omega' \in A$ whenever $\omega \in A$.

For $A \subseteq \Omega$ and $F \subseteq E$, we say that A *is defined on* F if $A = C \times \{0, 1\}^{E \setminus F}$ for some $C \subseteq \{0, 1\}^F$. We refer to F as the 'base' of the event A. If A is defined on F, we need only know the $\omega(e), e \in F$, to determine whether or not A occurs.

2.5 Theorem [85]. *Let $F \subseteq E$, and let A and B be increasing subsets of Ω such that A is defined on F, and B is defined on $E \setminus F$. Then*

$$\mathbb{P}(T \in A \mid T \in B) \le \mathbb{P}(T \in A).$$

Theorem 2.1 is retrieved by setting $A = \{\omega \in \Omega : \omega(f) = 1\}$ and $B = \{\omega \in \Omega : \omega(g) = 1\}$. The original proof of Theorem 2.5 is set in the context of matroid theory, and a further proof may be found in [32].

Whereas 'positive association' is well developed and understood as a technique for studying interacting systems, 'negative association' possesses some inherent difficulties. See [198] for further discussion.

2.2 Wilson's algorithm

There are various ways to generate a uniform spanning tree (UST) of the graph G. The following method, called *Wilson's algorithm* [240], highlights the close relationship between UST and random walk.

Take $G = (V, E)$ to be a finite connected graph. We shall perform random walks on G subject to a process of so-called *loop-erasure* that we describe next.[1] Let $\mathcal{W} = (w_0, w_1, \ldots, w_k)$ be a walk on G, which is to say that $w_i \sim w_{i+1}$ for $0 \leq i < k$ (note that the walk may have self-intersections). From $\mathcal{W}$, we construct a non-self-intersecting sub-walk, denoted $\mathrm{LE}(\mathcal{W})$, by the removal of loops as they occur. More precisely, let

$$J = \min\{j \geq 1 : w_j = w_i \text{ for some } i < j\},$$

and let I be the unique value of i satisfying $I < J$ and $w_I = w_J$. Let $\mathcal{W}' = (w_0, w_1, \ldots, w_I, w_{J+1}, \ldots, w_k)$ be the sub-walk of $\mathcal{W}$ obtained through the removal of the cycle $(w_I, w_{I+1}, \ldots, w_J)$. This operation of single-loop-removal is iterated until no loops remain, and we denote by $\mathrm{LE}(\mathcal{W})$ the surviving path from w_0 to w_k.

Wilson's algorithm is presented next. First, let $V = (v_1, v_2, \ldots, v_n)$ be an arbitrary but fixed ordering of the vertex-set.

1. Perform a random walk on G beginning at v_{i_1} with $i_1 = 1$, and stopped at the first time it visits v_n. The outcome is a walk $W_1 = (u_1 = v_1, u_2, \ldots, u_r = v_n)$.

2. From W_1, we obtain the loop-erased path $\mathrm{LE}(W_1)$, joining v_1 to v_n and containing no loops.[2] Set $T_1 = \mathrm{LE}(W_1)$.

3. Find the earliest vertex, v_{i_2} say, of V not belonging to T_1, and perform a random walk beginning at v_{i_2}, and stopped at the first moment it hits some vertex of T_1. Call the resulting walk W_2, and loop-erase W_2 to obtain some non-self-intersecting path $\mathrm{LE}(W_2)$ from v_{i_2} to T_1. Set $T_2 = T_1 \cup \mathrm{LE}(W_2)$, the union of two edge-disjoint paths.

---

[1] Graph theorists might prefer to call this *cycle-erasure*.

[2] If we run a random walk and then erase its loops, the outcome is called *loop-erased random walk*, often abbreviated to LERW.

4. Iterate the above process, by running and loop-erasing a random walk from a new vertex $v_{i_{j+1}} \notin T_j$ until it strikes the set T_j previously constructed.

5. Stop when all vertices have been visited, and set $T = T_N$, the final value of the T_j.

Each stage of the above algorithm results in a sub-tree of G. The final such sub-tree T is spanning since, by assumption, it contains every vertex of V.

2.6 Theorem [240]. *The graph T is a uniform spanning tree of G.*

Note that the initial ordering of V plays no role in the law of T.

There are of course other ways of generating a UST on G, and we mention the well known Aldous–Broder algorithm, [17, 53], that proceeds as follows. Choose a vertex r of G and perform a random walk on G, starting at r, until every vertex has been visited. For $w \in V$, $w \neq r$, let $[v, w\rangle$ be the directed edge that was traversed by the walk on its first visit to w. The edges thus obtained, when undirected, constitute a uniform spanning tree. The Aldous–Broder algorithm is closely related to Wilson's algorithm via a certain reversal of time, see [203] and Exercise 2.1.

We present the proof of Theorem 2.6 in a more general setting than UST. Heavy use will be made of [181] and the concept of 'cycle popping' introduced in the original paper [240] of David Wilson. Of considerable interest is an analysis of the run-time of Wilson's algorithm, see [203].

Consider an irreducible Markov chain with transition matrix P on the finite state space S. With this chain we may associate a directed graph $H = (S, F)$ much as in Section 1.1. The graph H has vertex-set S, and edge-set $F = \{[x, y\rangle : p_{x,y} > 0\}$. We refer to x (respectively, y) as the *head* (respectively, *tail*) of the (directed) edge $e = [x, y\rangle$, written $x = e_-$, $y = e_+$. Since the chain is irreducible, H is connected in the sense that, for all $x, y \in S$, there exists a directed path from x to y.

Let $r \in S$ be a distinguished vertex called the *root*. A *spanning arborescence* of H with root r is a subgraph A with the following properties:

(a) each vertex of S apart from r is the head of a unique edge of A,

(b) the root r is the head of no edge of A,

(c) A possesses no (directed) cycles.

Let Σ_r be the set of all spanning arborescences with root r, and $\Sigma = \bigcup_{r \in S} \Sigma_r$. A spanning arborescence is specified by its edge-set.

It is easily seen that there exists a unique (directed) path in the spanning arborescence A joining any given vertex x to the root. To the spanning arborescence A we assign the weight

(2.7) $$\alpha(A) = \prod_{e \in A} p_{e_-, e_+},$$

and we shall describe a randomized algorithm that selects a given spanning arborescence A with probability proportional to $\alpha(A)$. Since $\alpha(A)$ contains no diagonal element $p_{z,z}$ of P, and each x ($\neq r$) is the head of a unique edge of A, we may assume that $p_{z,z} = 0$ for all $z \in S$.

Let $r \in S$. Wilson's algorithm is easily adapted in order to sample from Σ_r. Let $v_1, v_2, \ldots, v_{n-1}$ be an ordering of $S \setminus \{r\}$.

1. Let $\sigma_0 = \{r\}$.
2. Sample a Markov chain with transition matrix P beginning at v_{i_1} with $i_1 = 1$, and stopped at the first time it hits σ_0. The outcome is a (directed) walk $W_1 = (u_1 = v_1, u_2, \ldots, u_k, r)$. From W_1, we obtain the loop-erased path $\sigma_1 = \mathrm{LE}(W_1)$, joining v_1 to r and containing no loops.
3. Find the earliest vertex, v_{i_2} say, of S not belonging to σ_1, and sample a Markov chain beginning at v_{i_2}, and stopped at the first moment it hits some vertex of σ_1. Call the resulting walk W_2, and loop-erase it to obtain some non-self-intersecting path $\mathrm{LE}(W_2)$ from v_{i_2} to σ_1. Set $\sigma_2 = \sigma_1 \cup \mathrm{LE}(W_2)$, the union of σ_1 and the directed path $\mathrm{LE}(W_2)$.
4. Iterate the above process, by loop-erasing the trajectory of a Markov chain starting at a new vertex $v_{i_{j+1}} \notin \sigma_j$ until it strikes the graph σ_j previously constructed.
5. Stop when all vertices have been visited, and set $\sigma = \sigma_N$, the final value of the σ_j.

2.8 Theorem [240]. *The graph σ is a spanning arborescence with root r, and*

$$\mathbb{P}(\sigma = A) \propto \alpha(A), \qquad A \in \Sigma_r.$$

Since S is finite and the chain is assumed irreducible, there exists a unique stationary distribution $\pi = (\pi_s : s \in S)$. Suppose that the chain is reversible with respect to π in that

$$\pi_x p_{x,y} = \pi_y p_{y,x}, \qquad x, y \in S.$$

As in Section 1.1, to each edge $e = [x, y\rangle$ we may allocate the weight $w(e) = \pi_x p_{x,y}$, noting that the edges $[x, y\rangle$ and $[y, x\rangle$ have equal weight. Let A be a spanning arborescence with root r. Since each vertex of H other than the root is the head of a unique edge of the spanning arborescence A, we have by (2.7) that

$$\alpha(A) = \frac{\prod_{e \in A} \pi_{e_-} p_{e_-, e_+}}{\prod_{x \in S,\, x \neq r} \pi_x} = CW(A), \qquad A \in \Sigma_r,$$

where $C = C_r$ and

(2.9)
$$W(A) = \prod_{e \in A} w(e).$$

Therefore, for a given root r, the weight functions α and W generate the same probability measure on Σ_r.

We shall see that the UST measure on $G = (V, E)$ arises through a consideration of the random walk on G. This has transition matrix given by

$$p_{x,y} = \begin{cases} \dfrac{1}{\deg(x)} & \text{if } x \sim y, \\ 0 & \text{otherwise,} \end{cases}$$

and stationary distribution

$$\pi_x = \frac{\deg(x)}{2|E|}, \qquad x \in V.$$

Let $H = (V, F)$ be the graph obtained from G by replacing each edge by a pair of edges with opposite orientations. Now, $w(e) = \pi_{e_-} p_{e_-,e_+}$ is independent of $e \in F$, so that $W(A)$ is a constant function. By Theorem 2.8 and the observation following (2.9), Wilson's algorithm generates a uniform random spanning arborescence σ of H, with given root. When we neglect the orientations of the edges of σ, and also the identity of the root, σ is transformed into a uniform spanning tree of G.

The remainder of this section is devoted to a proof of Theorem 2.8, and it uses the beautiful construction presented in [240]. We prepare for the proof as follows.

For each $x \in S \setminus \{r\}$, we provide ourselves in advance with an infinite set of 'moves' from x. Let $M_x(i)$, $i \geq 1$, $x \in S \setminus \{r\}$, be independent random variables with laws

$$\mathbb{P}(M_x(i) = y) = p_{x,y}, \qquad y \in S.$$

For each x, we organize the $M_x(i)$ into an ordered 'stack'. We think of an element $M_x(i)$ as having 'colour' i, where the colours indexed by i are distinct. The root r is given an empty stack. At stages of the following construction, we shall discard elements of stacks in order of increasing colour, and we shall call the set of uppermost elements of the stacks the 'visible moves'.

The visible moves generate a directed subgraph of H termed the 'visible graph'. There will generally be directed cycles in the visible graph, and we shall remove such cycles one by one. Whenever we decide to remove a cycle, the corresponding visible moves are removed from the stacks, and a new set of moves beneath is revealed. The visible graph thus changes, and a second cycle may be removed. This process may be iterated until the earliest time, N say, at which the visible graph contains no cycle, which is to say that the visible graph is a spanning arborescence σ with root r. If $N < \infty$, we terminate the procedure and 'output' σ. The removal of a cycle is called 'cycle popping'. It would seem that the value of N and the output σ will depend on the order in which we decide to pop cycles, but the converse turns out to be the case.

The following lemma holds 'pointwise': it contains no statement involving probabilities.

2.10 Lemma. *The order of cycle popping is irrelevant to the outcome, in that*:

 either $N = \infty$ *for all orderings of cycle popping,*

  or the total number N of popped cycles, and the output σ, are independent of the order of popping.

Proof. A *coloured cycle* is a set $M_{x_j}(i_j)$, $j = 1, 2, \ldots, J$, of moves, indexed by vertices x_j and colours i_j, with the property that they form a cycle of the graph H. A coloured cycle C is called *poppable* if there exists a sequence $C_1, C_2, \ldots, C_n = C$ of coloured cycles that may be popped in sequence. We claim the following for any cycle-popping algorithm. If the algorithm terminates in finite time, then all poppable cycles are popped, and no others. The lemma follows from this claim.

Let C be a poppable coloured cycle, and let $C_1, C_2, \ldots, C_n = C$ be as above. It suffices to show the following. Let $C' \neq C_1$ be a poppable cycle every move of which has colour 1, and suppose we pop C' at the first stage, rather than C_1. *Then C is still poppable after the removal of C'.*

Let $V(D)$ denote the vertex-set of a coloured cycle D. The italicized claim is evident if $V(C') \cap V(C_k) = \varnothing$ for $k = 1, 2, \ldots, n$. Suppose on the contrary that $V(C') \cap V(C_k) \neq \varnothing$ for some k, and let K be the earliest such k. Let $x \in V(C') \cap V(C_K)$. Since $x \notin V(C_k)$ for $k < K$, the visible move at x has colour 1 even after the popping of $C_1, C_2, \ldots, C_{K-1}$. Therefore, the edge of C_K with head x has the same tail, y say, as that of C' with head x. This argument may be applied to y also, and then to all vertices of C_K in order. In conclusion, C_K has colour 1, and $C' = C_K$.

Were we to decide to pop C' first, then we may choose to pop in the sequence $C_K [= C'], C_1, C_2, C_3, \ldots, C_{K-1}, C_{K+1}, \ldots, C_n = C$, and the claim has been shown. $\square$

Proof of Theorem 2.8. It is clear by construction that Wilson's algorithm terminates after finite time, with probability 1. It proceeds by popping cycles, and so, by Lemma 2.10, $N < \infty$ almost surely, and the output σ is independent of the choices available in its implementation.

We show next that σ has the required law. We may think of the stacks as generating a pair $(\mathbf{C}, \sigma)$, where $\mathbf{C} = (C_1, C_2, \ldots, C_J)$ is the ordered set of coloured cycles that are popped by Wilson's algorithm, and σ is the spanning arborescence thus revealed. Note that the colours of the moves of σ are determined by knowledge of $\mathbf{C}$. Let $\mathcal{C}$ be the set of all sequences $\mathbf{C}$ that may occur, and Π the set of all possible pairs $(\mathbf{C}, \sigma)$. Certainly $\Pi = \mathcal{C} \times \Sigma_r$, since knowledge of $\mathbf{C}$ imparts no information about σ.

The spanning arborescence σ contains exactly one coloured move in each stack (other than the empty stack at the root). Stacked above this move are a number of coloured moves, each of which belongs to exactly one of the popped cycles C_j. Therefore, the law of $(\mathbf{C}, \sigma)$ is given by the probability that the coloured moves in and above σ are given appropriately. That is,

$$\mathbb{P}\big((\mathbf{C}, \sigma) = (\mathbf{c}, A)\big) = \left(\prod_{c \in \mathbf{c}} \prod_{e \in c} p_{e_-, e_+}\right) \alpha(A), \qquad \mathbf{c} \in \mathcal{C}, \ A \in \Sigma_r.$$

Since this factorizes in the form $f(\mathbf{c})g(A)$, the random variables $\mathbf{C}$ and σ are independent, and $\mathbb{P}(\sigma = A)$ is proportional to $\alpha(A)$ as required. □

2.3 Weak limits on lattices

This section is devoted to the uniform-spanning-tree measure on the d-dimensional cubic lattice $\mathbb{L}^d = (\mathbb{Z}^d, \mathbb{E}^d)$ with $d \geq 2$. The UST is not usually defined directly on this graph, since it is infinite. It is defined instead on a finite subgraph Λ, and the limit is taken as $\Lambda \uparrow \mathbb{Z}^d$. Thus, we are led to study limits of probability measures, and to appeal to the important technique known as 'weak convergence'. This technique plays a major role in much of the work described in this volume. Readers in need of a good book on the topic are referred to the classic texts [39, 73]. They may in addition find the notes at the end of this section to be useful.

Let μ_n be the UST measure on the box $\Lambda(n) = [-n, n]^d$ of the lattice $\mathbb{L}^d$. Here and later, we shall consider the μ_n as probability measures on the measurable pair $(\Omega, \mathcal{F})$ comprising: the sample space $\Omega = \{0, 1\}^{\mathbb{E}^d}$, and the σ-algebra $\mathcal{F}$ of Ω generated by the cylinder sets. Elements of Ω are written $\omega = (\omega(e) : e \in \mathbb{E}^d)$.

2.11 Theorem [195]. *The weak limit* $\mu = \lim_{n \to \infty} \mu_n$ *exists and is a translation-invariant and ergodic probability measure. It is supported on the set of forests of $\mathbb{L}^d$ with no bounded component.*

Here is some further explanation of the language of this theorem. So-called 'ergodic theory' is concerned with certain actions on the sample space. Let $x \in \mathbb{Z}^d$. The function π_x acts on $\mathbb{Z}^d$ by $\pi_x(y) = x + y$; it is the *translation* by x, and it is a graph automorphism in that $\langle u, v \rangle \in \mathbb{E}^d$ if and only if $\langle \pi_x(u), \pi_x(v) \rangle \in \mathbb{E}^d$. The translation π_x acts on $\mathbb{E}^d$ by $\pi_x(\langle u, v \rangle) = (\pi_x(u), \pi_x(v))$, and on Ω by $\pi_x(\omega) = (\omega(\pi_{-x}(e)) : e \in \mathbb{E}^d)$.

An event $A \in \mathcal{F}$ is called *shift-invariant* if $A = \{\pi_x(\omega) : \omega \in A\}$ for every $x \in \mathbb{Z}^d$. A probability measure ϕ on $(\Omega, \mathcal{F})$ is *ergodic* if every shift-invariant event A is such that $\phi(A)$ is either 0 or 1. The measure is said to be *supported* on the event F if $\phi(F) = 1$.

Since we are working with the σ-field of Ω generated by the cylinder events, it suffices for weak convergence that $\mu_n(B \subseteq T) \to \mu(B \subseteq T)$ for any finite set B of edges (see the notes at the end of this section, and Exercise 2.4). Note that the limit measure μ may place strictly positive probability on the set of forests with two or more components. By a mild extension of the proof of Theorem 2.11, we obtain that the limit measure μ is invariant under the action of any automorphism of the lattice $\mathbb{L}^d$.

Proof. Let F be a finite set of edges of $\mathbb{E}^d$. By the Rayleigh principle, Theorem 1.29 (as in the proof of Theorem 2.1, see Exercise 2.5),

$$(2.12) \qquad \mu_n(F \subseteq T) \geq \mu_{n+1}(F \subseteq T),$$

for all large n. Therefore, the limit

$$\mu(F \subseteq T) = \lim_{n \to \infty} \mu_n(F \subseteq T)$$

exists. The domain of μ may be extended to all cylinder events, by the inclusion–exclusion principle or otherwise (see Exercises 2.3–2.4), and this in turn specifies a unique probability measure μ on $(\Omega, \mathcal{F})$. Since no tree contains a cycle, and since each cycle is finite and there are countably many cycles in $\mathbb{L}^d$, μ has support in the set of forests. By a similar argument, these forests may be taken with no bounded component.

Let π be a translation of $\mathbb{Z}^2$, and let F be finite as above. Then

$$\mu(\pi F \subseteq T) = \lim_{n \to \infty} \mu_n(\pi F \subseteq T) = \lim_{n \to \infty} \mu_{\pi,n}(F \subseteq T),$$

where $\mu_{\pi,n}$ is the law of a UST on $\pi^{-1}\Lambda(n)$. There exists $r = r(\pi)$ such that $\Lambda(n - r) \subseteq \pi^{-1}\Lambda(n) \subseteq \Lambda(n + r)$ for all large n. By the Rayleigh principle again,

$$\mu_{n+r}(F \subseteq T) \leq \mu_{\pi,n}(F \subseteq T) \leq \mu_{n-r}(F \subseteq T)$$

for all large n. Therefore,

$$\lim_{n \to \infty} \mu_{\pi,n}(F \subseteq T) = \mu(F \subseteq T),$$

whence the translation-invariance of μ. The proof of ergodicity is omitted, and may be found in [195].   □

This leads immediately to the question of whether or not the support of μ is the set of spanning trees of $\mathbb{L}^d$. The proof of the following is omitted.

2.13 Theorem [195]. *The limit measure μ is supported on the set of spanning trees of $\mathbb{L}^d$ if and only if $d \leq 4$.*

The measure μ may be termed 'free UST measure', where the word 'free' refers to the fact that no further assumption is made about the boundary $\partial \Lambda(n)$. There is another boundary condition giving rise to the so-called 'wired UST measure': we identify as a single vertex all vertices not in $\Lambda(n-1)$, and choose a spanning tree uniformly at random from the resulting (finite) graph. We can pass to the limit as $n \to \infty$ in very much the same way as before, with inequality (2.12) reversed. It turns out that the free and wired measures are identical on $\mathbb{L}^d$ for all d. The key reason is that $\mathbb{L}^d$ is a so-called *amenable* graph, which amounts in this context to saying that the boundary/volume approaches zero in the limit of large boxes,

$$\frac{|\partial \Lambda(n)|}{|\Lambda(n)|} \sim c \frac{n^{d-1}}{n^d} \to 0 \qquad \text{as } n \to \infty.$$

See Exercise 2.9 and [32, 181, 195, 196] for further details and discussion.

This section closes with a brief note about weak convergence, for more details of which the reader is referred to the books [39, 73]. Let $E = \{e_i : 1 \leq i < \infty\}$ be a countably infinite set. The product space $\Omega = \{0, 1\}^E$ may be viewed as the product of copies of the discrete topological space $\{0, 1\}$ and, as such, Ω is compact, and is metrisable by

$$\delta(\omega, \omega') = \sum_{i=1}^{\infty} 2^{-i} |\omega(e_i) - \omega'(e_i)|, \qquad \omega, \omega' \in \Omega.$$

A subset C of Ω is called a *(finite-dimensional) cylinder event* (or, simply, a cylinder) if there exists a finite subset $F \subseteq E$ such that: $\omega \in C$ if and only if $\omega' \in C$ for all ω' equal to ω on F. The *product σ-algebra* $\mathcal{F}$ of Ω is the σ-algebra generated by the cylinders. The *Borel σ-algebra* $\mathcal{B}$ of Ω is defined as the minimal σ-algebra containing the open sets. It is standard that $\mathcal{B}$ is generated by the cylinders, and therefore $\mathcal{F} = \mathcal{B}$ in the current setting. We note that every cylinder is both open and closed in the product topology.

Let $(\mu_n : n \geq 1)$ and μ be probability measures on $(\Omega, \mathcal{F})$. We say that μ_n converges *weakly* to μ, written $\mu_n \Rightarrow \mu$, if

$$\mu_n(f) \to \mu(f) \qquad \text{as } n \to \infty,$$

for all bounded continuous functions $f : \Omega \to \mathbb{R}$. (Here and later, $P(f)$ denotes the expectation of the function f under the measure P.) Several other definitions of weak convergence are possible, and the so-called 'portmanteau theorem' asserts that certain of these are equivalent. In particular, the weak convergence of μ_n to μ is equivalent to each of the two following statements:

(a) $\lim \sup_{n \to \infty} \mu_n(C) \leq \mu(C)$ for all closed events C,

(b) $\lim \inf_{n \to \infty} \mu_n(C) \geq \mu(C)$ for all open events C.

The matter is simpler in the current setting: since the cylinder events are both open and closed, and they generate $\mathcal{F}$, it is necessary and sufficient for weak convergence that

(c) $\lim_{n \to \infty} \mu_n(C) = \mu(C)$ for all cylinders C.

The following is useful for the construction of infinite-volume measures in the theory of interacting systems. Since Ω is compact, every family of probability measures on $(\Omega, \mathcal{F})$ is relatively compact. That is to say, for any such family $\Pi = (\mu_i : i \in I)$, every sequence $(\mu_{n_k} : k \geq 1)$ in Π possesses a weakly convergent subsequence. Suppose now that $(\mu_n : n \geq 1)$ is a sequence of probability measures on $(\Omega, \mathcal{F})$. If the limits $\lim_{n \to \infty} \mu_n(C)$ exists for every cylinder C, then it is necessarily the case that $\mu := \lim_{n \to \infty} \mu_n$ exists and is a probability measure. We shall see in Exercises 2.3–2.4 that this holds if and only if $\lim_{n \to \infty} \mu_n(C)$ exists for all *increasing* cylinders C. This justifies the argument of the proof of Theorem 2.11.

2.4 Uniform forest

We saw in Theorems 2.1 and 2.5 that the UST has a property of negative association. There is evidence that certain related measures have such a property also, but such claims have resisted proof.

Let $G = (V, E)$ be a finite graph, which we may as well assume to be connected. Write $\mathcal{F}$ for the set of forests of G (that is, subsets $H \subseteq E$ containing no cycle), and $\mathcal{C}$ for the set of connected subgraphs of G (that is, subsets $H \subseteq E$ such that (V, H) is connected). Let F be a uniformly chosen member of $\mathcal{F}$, and C a uniformly chosen member of $\mathcal{C}$. We refer to F and C as a *uniform forest* (UF) and a *uniform connected subgraph* (USC), respectively.

2.14 Conjecture. *For $f, g \in E$, $f \neq g$, the UF and USC satisfy:*

$$(2.15) \qquad \mathbb{P}(f \in F \mid g \in F) \leq \mathbb{P}(f \in F),$$

$$(2.16) \qquad \mathbb{P}(f \in C \mid g \in C) \leq \mathbb{P}(f \in C).$$

This is a special case of a more general conjecture for random-cluster measures indexed by the parameters $p \in (0, 1)$ and $q \in (0, 1)$. See Section 8.4.

We may further ask whether UF and USC might satisfy the stronger conclusion of Theorem 2.5. As positive evidence of Conjecture 2.14, we cite the computer-aided proof of [124] that the UF on any graph with eight or fewer vertices (or nine vertices and eighteen or fewer edges) satisfies (2.15).

Negative association presents difficulties that are absent from the better established theory of positive association (see Sections 4.1–4.2). There is an analogy with the concept of social (dis)agreement. Within a family or population, there may be a limited number of outcomes of consensus; there are generally many more outcomes of failure of consensus. Nevertheless, probabilists have made progress in developing systematic approaches to negative association, see for example [146, 198].

2.5 Schramm–Löwner evolutions

There is a beautiful result of Lawler, Schramm, and Werner [164] concerning the limiting LERW (loop-erased random walk) and UST measures on $\mathbb{L}^2$. This cannot be described without a detour into the theory of Schramm–Löwner evolutions (SLE).[3]

The theory of SLE is a major piece of contemporary mathematics which promises to explain phase transitions in an important class of two-dimensional disordered systems, and to help bridge the gap between probability theory and conformal field theory. It plays a key role in the study of critical percolation (see Chapter 5), and also of the critical random-cluster and Ising models, [224, 225]. In addition, it has provided complete explanations of conjectures made by mathematicians and physicists concerning the intersection exponents and fractionality of frontier of two-dimensional Brownian motion, [160, 161, 162]. The purposes of the current section are to give a brief non-technical introduction to SLE, and to indicate its relevance to the scaling limits of LERW and UST.

Let $\mathbb{H} = (-\infty, \infty) \times (0, \infty)$ be the upper half-plane of $\mathbb{R}^2$, with closure $\overline{\mathbb{H}}$, viewed as subsets of the complex plane. Consider the (Löwner) ordinary differential equation

$$\frac{d}{dt} g_t(z) = \frac{2}{g_t(z) - b(t)}, \qquad z \in \overline{\mathbb{H}} \setminus \{0\},$$

[3]SLE was originally an abbreviation for stochastic Löwner evolution, but is now regarded as named after Oded Schramm in recognition of his work reported in [215].

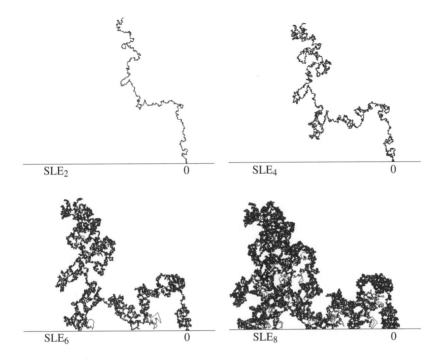

Figure 2.1. Simulations of the traces of chordal SLE_κ for $\kappa = 2, 4, 6, 8$. The four pictures are generated from the same Brownian driving path.

subject to the boundary condition $g_0(z) = z$, where $t \in [0, \infty)$, and $b : \mathbb{R} \to \mathbb{R}$ is termed the 'driving function'. Randomness is injected into this formula through setting $b(t) = B_{\kappa t}$ where $\kappa > 0$ and $(B_t : t \geq 0)$ is a standard Brownian motion.[4] The solution exists when $g_t(z)$ is bounded away from $B_{\kappa t}$. More specifically, for $z \in \overline{\mathbb{H}}$, let τ_z be the infimum of all times τ such that 0 is a limit point of $g_s(z) - B_{\kappa s}$ in the limit as $s \uparrow \tau$. We let

$$H_t = \{z \in \mathbb{H} : \tau_z > t\}, \qquad K_t = \{z \in \overline{\mathbb{H}} : \tau_z \leq t\},$$

so that H_t is open, and K_t is compact. It may now be seen that g_t is a conformal homeomorphism from H_t to $\mathbb{H}$. There exists a random curve $\gamma : [0, \infty) \to \overline{\mathbb{H}}$, called the *trace* of the process, such that $\mathbb{H} \setminus K_t$ is the unbounded component of $\mathbb{H} \setminus \gamma[0, t]$. The trace γ satisfies $\gamma(0) = 0$ and $\gamma(t) \to \infty$ as $t \to \infty$. (See the illustrations of Figure 2.1.)

---

[4]An interesting and topical account of the history and practice of Brownian motion may be found at [75].

We call $(g_t : t \geq 0)$ a *Schramm–Löwner evolution* (SLE) with parameter κ, written SLE_κ, and the K_t are called the *hulls* of the process. There is good reason to believe that the family $K = (K_t : t \geq 0)$ provides the correct scaling limits for a variety of random spatial processes, with the value of κ depending on the process in question. General properties of SLE_κ, viewed as a function of κ, have been studied in [207, 235, 236], and a beautiful theory has emerged. For example, the hulls K form (almost surely) a simple path if and only if $\kappa \leq 4$. If $\kappa > 8$, the trace of SLE_κ is (almost surely) a space-filling curve.

The above SLE process is termed 'chordal'. In another version, called 'radial' SLE, the upper half-plane $\mathbb{H}$ is replaced by the unit disc $\mathbb{U}$, and a different differential equation is satisfied. Let $\partial \mathbb{U}$ denote the boundary of $\mathbb{U}$. The corresponding curve γ satisfies $\gamma(t) \to 0$ as $t \to \infty$, and $\gamma(0) \in \partial \mathbb{U}$, say $\gamma(0)$ is uniformly distributed on $\partial \mathbb{U}$. Both chordal and radial SLE may be defined on an arbitrary simply connected domain D with a smooth boundary, by applying a suitable conformal map ϕ from either $\mathbb{H}$ or $\mathbb{U}$ to D.

It is believed that many discrete models in two dimensions, when at their critical points, converge in the limit of small mesh-size to an SLE_κ with κ chosen appropriately. Oded Schramm [215, 216] identified the correct values of κ for several different processes, and indicated that percolation has scaling limit SLE_6. Full rigorous proofs are not yet known even for general percolation models. For the special but presumably representative case of site percolation on the triangular lattice $\mathbb{T}$, Smirnov [222, 223] proved the very remarkable result that the crossing probabilities of re-scaled regions of $\mathbb{R}^2$ satisfy Cardy's formula (see Section 5.6), and he indicated the route to the full SLE_6 limit. See [56, 57, 58, 236] for more recent work on percolation, and [224, 225] for progress on the SLE limits of the critical random-cluster and Ising models in two dimensions.

This chapter closes with a brief summary of the results of [164] concerning SLE limits for loop-erased random walk (LERW) and uniform spanning tree (UST) on the square lattice $\mathbb{L}^2$. We saw earlier in this chapter that there is a very close relationship between LERW and UST on a finite connected graph G. For example, the unique path joining vertices u and v in a UST of G has the law of a LERW from u to v (see [195] and the description of Wilson's algorithm). See Figure 2.2.

Let D be a bounded simply connected subset of $\mathbb{C}$ with a smooth boundary ∂D and such that 0 lies in its interior. As remarked above, we may define radial SLE_2 on D, and we write ν for its law. Let $\delta > 0$, and let μ_δ be the law of LERW on the re-scaled lattice $\delta \mathbb{Z}^2$, starting at 0 and stopped when it first hits ∂D.

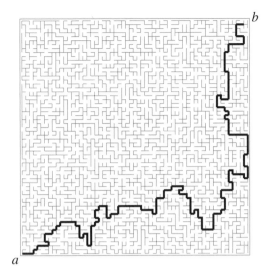

Figure 2.2. A uniform spanning tree (UST) on a large square of the square lattice. It contains a unique path between any two vertices a, b, and this has the law of a loop-erased random walk (LERW) between a and b.

For two parametrizable curves β, γ in $\mathbb{C}$, we define the distance between them by

$$\rho(\beta, \gamma) = \inf \left[\sup_{t \in [0,1]} |\widehat{\beta}(t) - \widehat{\gamma}(t)| \right],$$

where the infimum is over all parametrizations $\widehat{\beta}$ and $\widehat{\gamma}$ of the curves (see [8]). The distance function ρ generates a topology on the space of parametrizable curves, and hence a notion of weak convergence, denoted '$\Rightarrow$'.

2.17 Theorem [164]. *We have that* $\mu_\delta \Rightarrow \nu$ *as* $\delta \to 0$.

We turn to the convergence of UST to SLE_8, and begin with a discussion of mixed boundary conditions. Let D be a bounded simply connected domain of $\mathbb{C}$ with a smooth (C^1) boundary curve ∂D. For distinct points $a, b \in \partial D$, we write α (respectively, β) for the arc of ∂D going clockwise from a to b (respectively, b to a). Let $\delta > 0$ and let G_δ be a connected graph that approximates to that part of $\delta \mathbb{Z}^2$ lying inside D. We shall construct a UST of G_δ with mixed boundary conditions, namely a free boundary near α and a wired boundary near β. To each tree T of G_δ there corresponds a dual

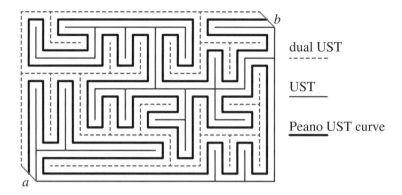

Figure 2.3. An illustration of the Peano UST path lying between a tree
and its dual. The thinner continuous line depicts the UST, and the dashed
line its dual tree. The thicker line is the Peano UST path.

tree T^{d} on the dual[5] graph G_δ^{d}, namely the tree comprising edges of G_δ^{d} that
do not intersect those of T. Since G_δ has mixed boundary conditions, so
does its dual G_δ^{d}. With G_δ and G_δ^{d} drawn together, there is a simple path
$\pi(T, T^{\mathrm{d}})$ that winds between T and T^{d}. Let Π be the path thus constructed
between the UST on G_δ and its dual tree. The construction of this 'Peano
UST path' is illustrated in Figures 2.3 and 2.4.

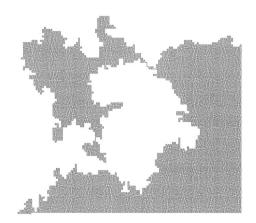

Figure 2.4. An initial segment of the Peano path constructed from a
UST on a large rectangle with mixed boundary conditions.

[5]This is the planar duality of graph theory, see page 41.

2.18 Theorem [164]. *The law of* Π *converges as* $\delta \to 0$ *to that of the image of chordal* SLE_8 *under any conformal map from* $\mathbb{H}$ *to* D *mapping* 0 *to* a *and* ∞ *to* b.

2.6 Exercises

2.1 [17, 53] *Aldous–Broder algorithm.* Let $G = (V, E)$ be a finite connected graph, and pick a root $r \in V$. Perform a random walk on G starting from r. For each $v \in V$, $v \neq r$, let e_v be the edge traversed by the random walk just before it hits v for the first time, and let T be the tree $\bigcup_v e_v$ rooted at r. Show that T, when viewed as an unrooted tree, is a uniform spanning tree. It may be helpful to argue as follows.

(a) Consider a stationary simple random walk $(X_n : -\infty < n < \infty)$ on G, with distribution $\pi_v \propto \deg(v)$, the degree of v. Let T_i be the rooted tree obtained by the above procedure applied to the sub-walk $X_i, X_{i+1}, \dots$. Show that $T = (T_i : -\infty < i < \infty)$ is a stationary Markov chain with state space the set $\mathcal{R}$ of rooted spanning trees.

(b) Let $Q(t, t') = \mathbb{P}(T_0 = t' \mid T_1 = t)$, and let $d(t)$ be the degree of the root of $t \in \mathcal{R}$. Show that:

    (i) for given $t \in \mathcal{R}$, there are exactly $d(t)$ trees $t' \in \mathcal{R}$ with $Q(t, t') = 1/d(t)$, and $Q(t, t') = 0$ for all other t',

    (ii) for given $t' \in \mathcal{R}$, there are exactly $d(t')$ trees $t \in \mathcal{R}$ with $Q(t, t') = 1/d(t)$, and $Q(t, t') = 0$ for all other t.

(c) Show that

$$\sum_{t \in \mathcal{R}} d(t) Q(t, t') = d(t'), \qquad t' \in \mathcal{R},$$

and deduce that the stationary measure of T is proportional to $d(t)$.

(d) Let $r \in V$, and let t be a tree with root r. Show that $\mathbb{P}(T_0 = t \mid X_0 = r)$ is independent of the choice of t.

2.2 *Inclusion–exclusion principle.* Let F be a finite set, and let f, g be real-valued functions on the power-set of F. Show that

$$f(A) = \sum_{B \subseteq A} g(B), \qquad A \subseteq F,$$

if and only if

$$g(A) = \sum_{B \subseteq A} (-1)^{|A \setminus B|} f(B), \qquad A \subseteq F.$$

Show the corresponding fact with the two summations replaced by $\sum_{B \supseteq A}$ and the exponent $|A \setminus B|$ by $|B \setminus A|$.

2.3 Let $\Omega = \{0, 1\}^F$, where F is finite, and let $\mathbb{P}$ be a probability measure on Ω, and $A \subseteq \Omega$. Show that $\mathbb{P}(A)$ may be expressed as a linear combination of certain $\mathbb{P}(A_i)$, where the A_i are increasing events.

2.4 (continuation) Let $G = (V, E)$ be an infinite graph with finite vertex-degrees, and $\Omega = \{0, 1\}^E$, endowed with the product σ-field. An event A in the product σ-field of Ω is called a *cylinder event* if it has the form $A_F \times \{0, 1\}^{\overline{F}}$ for some $A_F \subseteq \{0, 1\}^F$ and some finite $F \subseteq E$. Show that a sequence (μ_n) of probability measures converges weakly if and only if $\mu_n(A)$ converges for every increasing cylinder event A.

2.5 Let $G = (V, E)$ be a connected subgraph of the finite connected graph G'. Let T and T' be uniform spanning trees on G and G' respectively. Show that, for any edge e of G, $\mathbb{P}(e \in T) \geq \mathbb{P}(e \in T')$.

More generally, let B be a subset of E, and show that $\mathbb{P}(B \subseteq T) \geq \mathbb{P}(B \subseteq T')$.

2.6 Let T_n be a UST of the lattice box $[-n, n]^d$ of $\mathbb{Z}^d$. Show that the limit $\lambda(e) = \lim_{n \to \infty} \mathbb{P}(e \in T_n)$ exists.

More generally, show that the weak limit of T_n exists as $n \to \infty$.

2.7 Adapt the conclusions of the last two examples to the 'wired' UST measure μ^{W} on $\mathbb{L}^d$.

2.8 Let $\mathcal{F}$ be the set of forests of $\mathbb{L}^d$ with no bounded component, and let μ be an automorphism-invariant probability measure with support $\mathcal{F}$. Show that the mean degree of every vertex is 2.

2.9 [195] Let A be an increasing cylinder event in $\{0, 1\}^{\mathbb{E}^d}$, where $\mathbb{E}^d$ denotes the edge-set of the hypercubic lattice $\mathbb{L}^d$. Using the Feder–Mihail Theorem 2.5 or otherwise, show that the free and wired UST measures on $\mathbb{L}^d$ satisfy $\mu^{\mathrm{f}}(A) \geq \mu^{\mathrm{W}}(A)$. Deduce by the last exercise and Strassen's theorem, or otherwise, that $\mu^{\mathrm{f}} = \mu^{\mathrm{W}}$.

2.10 Consider the square lattice $\mathbb{L}^2$ as an infinite electrical network with unit edge-resistances. Show that the effective resistance between two neighbouring vertices is 2.

2.11 Let $G = (V, E)$ be finite and connected, and let $W \subseteq V$. Let $\mathcal{F}_W$ be the set of forests of G comprising exactly $|W|$ trees with respective roots the members of W. Explain how Wilson's algorithm may be adapted to sample uniformly from $\mathcal{F}_W$.

3

Percolation and self-avoiding walk

The central feature of the percolation model is the phase transition.
The existence of the point of transition is proved by path-counting
and planar duality. Basic facts about self-avoiding walks, oriented
percolation, and the coupling of models are reviewed.

3.1 Percolation and phase transition

Percolation is the fundamental stochastic model for spatial disorder. In
its simplest form introduced in [52][1], it inhabits a (crystalline) lattice and
possesses the maximum of (statistical) independence. We shall consider
mostly percolation on the (hyper)cubic lattice $\mathbb{L}^d = (\mathbb{Z}^d, \mathbb{E}^d)$ in $d \geq 2$
dimensions, but much of the following may be adapted to an arbitrary lattice.

Percolation comes in two forms, 'bond' and 'site', and we concentrate
here on the bond model. Let $p \in [0, 1]$. Each edge $e \in \mathbb{E}^d$ is designated either *open* with probability p, or *closed* otherwise, different edges
receiving independent states. We think of an open edge as being open to
the passage of some material such as disease, liquid, or infection. Suppose
we remove all closed edges, and consider the remaining open subgraph of
the lattice. Percolation theory is concerned with the geometry of this open
graph. Of particular interest are such quantites as the size of the open cluster
C_x containing a given vertex x, and particularly the probability that C_x is
infinite.

The sample space is the set $\Omega = \{0, 1\}^{\mathbb{E}^d}$ of 0/1-vectors ω indexed
by the edge-set; here, 1 represents 'open', and 0 'closed'. As σ-field we
take that generated by the finite-dimensional cylinder sets, and the relevant
probability measure is product measure $\mathbb{P}_p$ with density p.

For $x, y \in \mathbb{Z}^d$, we write $x \leftrightarrow y$ if there exists an open path joining x and
y. The *open cluster* C_x at x is the set of all vertices reachable along open

---

[1] See also [241].

paths from the vertex x,

$$C_x = \{y \in \mathbb{Z}^d : x \leftrightarrow y\}.$$

The origin of $\mathbb{Z}^d$ is denoted 0, and we write $C = C_0$. The principal object of study is the *percolation probability* $\theta(p)$ given by

$$\theta(p) = \mathbb{P}_p(|C| = \infty).$$

The critical probability is defined as

(3.1) $$p_c = p_c(\mathbb{L}^d) = \sup\{p : \theta(p) = 0\}.$$

It is fairly clear (and will be spelled out in Section 3.3) that θ is non-decreasing in p, and thus

$$\theta(p) \begin{cases} = 0 & \text{if } p < p_c, \\ > 0 & \text{if } p > p_c. \end{cases}$$

It is fundamental that $0 < p_c < 1$, and we state this as a theorem. It is easy to see that $p_c = 1$ for the corresponding one-dimensional process.

3.2 Theorem. *For $d \geq 2$, we have that $0 < p_c < 1$.*

The inequalities may be strengthened using counts of self-avoiding walks, as in Theorem 3.12. It is an important open problem to prove the following conjecture. The conclusion is known only for $d = 2$ and $d \geq 19$.

3.3 Conjecture. *For $d \geq 2$, we have that $\theta(p_c) = 0$.*

It is the edges (or 'bonds') of the lattice that are declared open/closed above. If, instead, we designate the vertices (or 'sites') to be open/closed, the ensuing model is termed *site percolation*. Subject to minor changes, the theory of site percolation may be developed just as that of bond percolation.

Proof of Theorem 3.2. This proof introduces two basic methods, namely the counting of paths and the use of planar duality. We show first by counting paths that $p_c > 0$.

A *self-avoiding walk* (SAW) is a lattice path that visits no vertex more than once. Let σ_n be the number of SAWs with length n beginning at the origin, and let N_n be the number of such SAWs all of whose edges are open. Then

$$\theta(p) = \mathbb{P}_p(N_n \geq 1 \text{ for all } n \geq 1)$$
$$= \lim_{n \to \infty} \mathbb{P}_p(N_n \geq 1).$$

Now,

(3.4) $$\mathbb{P}_p(N_n \geq 1) \leq \mathbb{E}_p(N_n) = p^n \sigma_n.$$

A a crude upper bound for σ_n, we have that

$$(3.5) \qquad \sigma_n \leq 2d(2d-1)^{n-1}, \qquad n \geq 1,$$

since the first step of a SAW from the origin can be to any of its $2d$ neighbours, and there are no more than $2d-1$ choices for each subsequent step. Thus

$$\theta(p) \leq \lim_{n\to\infty} 2d(2d-1)^{n-1}p^n,$$

which equals 0 whenever $p(2d-1) < 1$. Therefore,

$$p_c \geq \frac{1}{2d-1}.$$

We turn now to the proof that $p_c < 1$. The first step is to observe that

$$(3.6) \qquad p_c(\mathbb{L}^d) \geq p_c(\mathbb{L}^{d+1}), \qquad d \geq 2.$$

This follows by the observation that $\mathbb{L}^d$ may be embedded in $\mathbb{L}^{d+1}$ in such a way that the origin lies in an infinite open cluster of $\mathbb{L}^{d+1}$ whenever it lies in an infinite open cluster of the smaller lattice $\mathbb{L}^d$. By (3.6), it suffices to show that

$$(3.7) \qquad p_c(\mathbb{L}^2) < 1,$$

and to this end we shall use a technique known as planar duality, which arises as follows.

Let G be a planar graph, drawn in the plane. The *planar dual* of G is the graph constructed in the following way. We place a vertex in every face of G (including the infinite face if it exists) and we join two such vertices by an edge if and only if the corresponding faces of G share a boundary edge. It is easy to see that the dual of the square lattice $\mathbb{L}^2$ is a copy of $\mathbb{L}^2$, and we refer therefore to the square lattice as being *self-dual*. See Figure 1.5.

There is a natural one–one correspondence between the edge-set of the dual lattice $\mathbb{L}_d^2$ and that of the primal $\mathbb{L}^2$, and this gives rise to a percolation model on $\mathbb{L}_d^2$ by: for an edge $e \in \mathbb{E}^2$ and it dual edge e_d, we declare e_d to be open if and only if e is open. As illustrated in Figure 3.1, each finite open cluster of $\mathbb{L}^2$ lies in the interior of a closed cycle of $\mathbb{L}_d^2$ lying 'just outside' the cluster.

We use a so-called Peierls argument[2] to obtain (3.7). Let M_n be the number of closed cycles of the dual lattice, having length n and containing

---

[2] This method was used by Peierls [194] to prove phase transition in the two-dimensional Ising model.

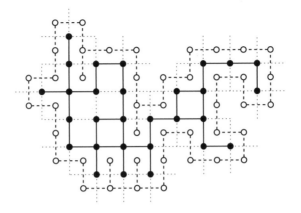

Figure 3.1. A finite open cluster of the primal lattice lies 'just inside' a closed cycle of the dual lattice.

0 in their interior. Note that $|C| < \infty$ if and only if $M_n \geq 1$ for some n. Therefore,

$$(3.8) \qquad 1 - \theta(p) = \mathbb{P}_p(|C| < \infty) = \mathbb{P}_p\left(\sum_n M_n \geq 1\right)$$

$$\leq \mathbb{E}_p\left(\sum_n M_n\right)$$

$$= \sum_{n=4}^{\infty} \mathbb{E}_p(M_n) \leq \sum_{n=4}^{\infty} (n4^n)(1-p)^n,$$

where we have used the facts that the shortest dual cycle containing 0 has length 4, and that the total number of dual cycles, having length n and surrounding the origin, is no greater than $n4^n$. The final sum may be made strictly smaller than 1 by choosing p sufficiently close to 1, say $p > 1 - \epsilon$ where $\epsilon > 0$. This implies that $p_c(\mathbb{L}^2) < 1 - \epsilon$ as required for (3.7). $\qquad\square$

3.2 Self-avoiding walks

How many self-avoiding walks of length n exist, starting from the origin? What is the 'shape' of a SAW chosen at random from this set? In particular, what can be said about the distance between its endpoints? These and related questions have attracted a great deal of attention since the publication in 1954 of the pioneering paper [130] of Hammersley and Morton, and never more so than in recent years. It is believed but not proved that a typical SAW on $\mathbb{L}^2$, starting at the origin, converges in a suitable manner as $n \to \infty$ to a $SLE_{8/3}$

curve, and the proof of this statement is an open problem of outstanding interest. See Section 2.5, in particular Figure 2.1, for an illustration of the geometry, and [183, 216, 224] for discussion and results.

The use of subadditivity was one of the several stimulating ideas of [130], and it has proved extremely fruitful in many contexts since. Consider the lattice $\mathbb{L}^d$, and let $\mathcal{S}_n$ be the set of SAWs with length n starting at the origin, and $\sigma_n = |\mathcal{S}_n|$ as before.

3.9 Lemma. *We have that* $\sigma_{m+n} \leq \sigma_m \sigma_n$, *for* $m, n \geq 0$.

Proof. Let π and π' be finite SAWs starting at the origin, and denote by $\pi * \pi'$ the walk obtained by following π from 0 to its other endpoint x, and then following the translated walk $\pi' + x$. Every $\nu \in \mathcal{S}_{m+n}$ may be written in a unique way as $\nu = \pi * \pi'$ for some $\pi \in \mathcal{S}_m$ and $\pi' \in \mathcal{S}_n$. The claim of the lemma follows. $\qquad\square$

3.10 Theorem [130]. *The limit* $\kappa = \lim_{n \to \infty} (\sigma_n)^{1/n}$ *exists and satisfies* $d \leq \kappa \leq 2d - 1$.

This is in essence a consequence of the 'sub-multiplicative' inequality of Lemma 3.9. The constant κ is called the *connective constant* of the lattice. The exact value of $\kappa = \kappa(\mathbb{L}^d)$ is unknown for every $d \geq 2$, see [141, Sect. 7.2, pp. 481–483]. On the other hand, the 'hexagonal' (or 'honeycomb') lattice (see Figure 1.5) has a special structure which has permitted a proof by Duminil-Copin and Smirnov [74] that its connective constant equals $\sqrt{2 + \sqrt{2}}$.

Proof. By Lemma 3.9, $x_m = \log \sigma_m$ satisfies the 'subadditive inequality'

$$(3.11) \qquad\qquad x_{m+n} \leq x_m + x_n.$$

The existence of the limit

$$\lambda = \lim_{n \to \infty} \{x_n / n\}$$

follows immediately (see Exercise 3.1), and

$$\lambda = \inf_m \{x_m / m\} \in [-\infty, \infty).$$

By (3.5), $\kappa = e^\lambda \leq 2d - 1$. Finally, σ_n is at least the number of 'stiff' walks every step of which is in the direction of an *increasing* coordinate. The number of such walks is d^n, and therefore $\kappa \geq d$. $\qquad\square$

The bounds of Theorem 3.2 may be improved as follows.

3.12 Theorem. *The critical probability of bond percolation on* $\mathbb{L}^d$, *with* $d \geq 2$, *satisfies*

$$\frac{1}{\kappa(d)} \leq p_c \leq 1 - \frac{1}{\kappa(2)},$$

where $\kappa(d)$ *denotes the connective constant of* $\mathbb{L}^d$.

Proof. As in (3.4),

$$\theta(p) \leq \lim_{n \to \infty} p^n \sigma_n.$$

Now, $\sigma_n = \kappa(d)^{(1+o(1))n}$, so that $\theta(p) = 0$ if $p\kappa(d) < 1$.

For the upper bound, we elaborate on the proof of the corresponding part of Theorem 3.2. Let F_m be the event that there exists a closed cycle of the dual lattice $\mathbb{L}_d^2$ containing the primal box $\Lambda(m) = [-m, m]^2$ in its interior, and let G_m be the event that all edges of $\Lambda(m)$ are open. These two events are independent, since they are defined in terms of disjoint sets of edges. As in (3.8),

$$(3.13) \qquad \mathbb{P}_p(F_m) \leq \mathbb{P}_p\left(\sum_{n=4m}^{\infty} M_n \geq 1 \right)$$

$$\leq \sum_{n=4m}^{\infty} n(1-p)^n \sigma_n.$$

Recall that $\sigma_n = \kappa(2)^{(1+o(1))n}$, and choose p such that $(1-p)\kappa(2) < 1$. By (3.13), we may find m such that $\mathbb{P}_p(F_m) < \frac{1}{2}$. Then,

$$\theta(p) \geq \mathbb{P}_p(\overline{F_m} \cap G_m) = \mathbb{P}_p(\overline{F_m})\mathbb{P}_p(G_m) \geq \tfrac{1}{2}\mathbb{P}_p(G_m) > 0.$$

The upper bound on p_c follows. □

There are some extraordinary conjectures concerning SAWs in two dimensions. We mention the conjecture that

$$\sigma_n \sim An^{11/32}\kappa^n \qquad \text{when } d = 2.$$

This is expected to hold for any lattice in two dimensions, with an appropriate choice of constant A depending on the choice of lattice. It is known in contrast that no polynomial correction is necessary when $d \geq 5$,

$$\sigma_n \sim A\kappa^n \qquad \text{when } d \geq 5,$$

for the cubic lattice at least. Related to the above conjecture is the belief that a random SAW of $\mathbb{Z}^2$, starting at the origin and of length n, converges weakly as $n \to \infty$ to $\text{SLE}_{8/3}$. See [183, 216, 224] for further details of these and other conjectures and results.

3.3 Coupled percolation

The use of coupling in probability theory goes back at least as far as the beautiful proof by Doeblin of the ergodic theorem for Markov chains, [71]. In percolation, we couple together the bond models with different values of p as follows. Let U_e, $e \in \mathbb{E}^d$, be independent random variables with the uniform distribution on $[0, 1]$. For $p \in [0, 1]$, let

$$\eta_p(e) = \begin{cases} 1 & \text{if } U_e < p, \\ 0 & \text{otherwise.} \end{cases}$$

Thus, the configuration η_p $(\in \Omega)$ has law $\mathbb{P}_p$, and in addition

$$\eta_p \le \eta_r \quad \text{if} \quad p \le r.$$

3.14 Theorem. *For any increasing non-negative random variable $f : \Omega \to \Omega$, the function $g(p) = \mathbb{P}_p(f)$ is non-decreasing.*

Proof. For $p \le r$, we have that $\eta_p \le \eta_r$, whence $f(\eta_p) \le f(\eta_r)$. Therefore,

$$g(p) = \mathbb{P}(f(\eta_p)) \le \mathbb{P}(f(\eta_r)) = g(r),$$

as required. □

3.4 Oriented percolation

The 'north–east' lattice $\vec{\mathbb{L}}^d$ is obtained by orienting each edge of $\mathbb{L}^d$ in the direction of increasing coordinate-value (see Figure 3.2 for a two-dimensional illustration). There are many parallels between results for oriented percolation and those for ordinary percolation; on the other hand, the corresponding proofs often differ, largely because the existence of one-way streets restricts the degree of spatial freedom of the traffic.

Let $p \in [0, 1]$. We declare an edge of $\vec{\mathbb{L}}^d$ to be *open* with probability p and otherwise *closed*. The states of different edges are taken to be independent. We supply fluid at the origin, and allow it to travel along open edges in the directions of their orientations only. Let $\vec{C}$ be the set of vertices that may be reached from the origin along open directed paths. The *percolation probability* is

(3.15) $$\vec{\theta}(p) = \mathbb{P}_p(|\vec{C}| = \infty),$$

and the critical probability $\vec{p}_c(d)$ by

(3.16) $$\vec{p}_c(d) = \sup\{p : \vec{\theta}(p) = 0\}.$$

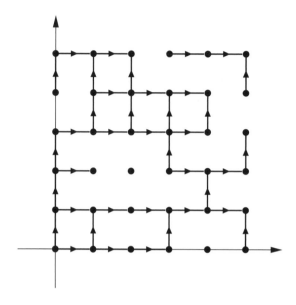

Figure 3.2. Part of the two-dimensional 'north–east' lattice in which each edge has been deleted with probability $1 - p$, independently of all other edges.

3.17 Theorem. *For $d \geq 2$, we have that $0 < \vec{p}_c(d) < 1$.*

Proof. Since an oriented path is also a path, it is immediate that $\vec{\theta}(p) \leq \theta(p)$, whence $\vec{p}_c(d) \geq p_c$. As in the proof of Theorem 3.2, it suffices for the converse to show that $\vec{p}_c = \vec{p}_c(2) < 1$.

Let $d = 2$. The cluster $\vec{C}$ comprises the endvertices of open edges that are oriented northwards/eastwards. Assume $|\vec{C}| < \infty$. We may draw a dual cycle Δ surrounding $\vec{C}$ in the manner illustrated in Figure 3.3. As we traverse Δ in the clockwise direction, we traverse dual edges each of which is oriented in one of the four compass directions. Any edge of Δ that is oriented either eastwards or southwards crosses a primal edge that is closed. Exactly one half of the edges of Δ are oriented thus, so that, as in (3.8),

$$\mathbb{P}_p(|\vec{C}| < \infty) \leq \sum_{n \geq 4} 4 \cdot 3^{n-2}(1 - p)^{\frac{1}{2}n-1}.$$

In particular, $\vec{\theta}(p) > 0$ if $1 - p$ is sufficiently small and positive. $\qquad\square$

The process is understood quite well when $d = 2$, see [77]. By looking at the set A_n of wet vertices on the diagonal $\{x \in \mathbb{Z}^2 : x_1 + x_2 = n\}$ of $\mathbb{L}^2$, we

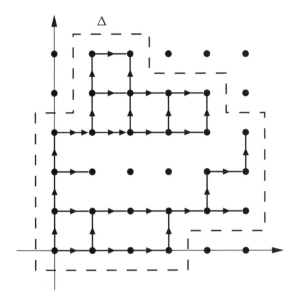

Figure 3.3.   As we trace the dual cycle Δ, we traverse edges exactly one half of which cross closed boundary edges of the cluster $\vec{C}$ at the origin.

may reformulate two-dimensional oriented percolation as a one-dimensional contact process in discrete time (see [167, Chap. 6]). It turns out that $\vec{p}_{c}(2)$ may be characterized in terms of the velocity of the rightwards edge of a contact process on $\mathbb{Z}$ whose initial distribution places infectives to the left of the origin and susceptibles to the right. With the support of arguments from branching processes and ordinary percolation, we may prove such results as the exponential decay of the cluster-size distribution when $p < \vec{p}_{c}(2)$, and its sub-exponential decay when $p > \vec{p}_{c}(2)$: there exist $\alpha(p), \beta(p) > 0$ such that

(3.18)
$$e^{-\alpha(p)\sqrt{n}} \leq \mathbb{P}_{p}(n \leq |\vec{C}| < \infty) \leq e^{-\beta(p)\sqrt{n}} \qquad \text{if } \vec{p}_{c}(2) < p < 1.$$

There is a close relationship between oriented percolation and the contact model (see Chapter 6), and methods developed for the latter model may often be applied to the former. It has been shown in particular that $\vec{\theta}(\vec{p}_{c}) = 0$ for general $d \geq 2$, see [112].

We close this section with an open problem of a different sort. Suppose that each edge of $\mathbb{L}^{2}$ is oriented in a random direction, horizontal edges being oriented eastwards with probability p and westwards otherwise, and vertical edges being oriented northwards with probability p and southwards otherwise. Let $\eta(p)$ be the probability that there exists an infinite oriented path starting at the origin. It is not hard to show that $\eta(\frac{1}{2}) = 0$ (see Exercise 3.9). We ask whether or not $\eta(p) > 0$ if $p \neq \frac{1}{2}$. Partial results in this direction may be found in [108], see also [174, 175].

3.5 Exercises

3.1 *Subadditive inequality.* Let $(x_n : n \geq 1)$ be a real sequence satisfying $x_{m+n} \leq x_m + x_n$ for $m, n \geq 1$. Show that the limit $\lambda = \lim_{n\to\infty}\{x_n/n\}$ exists and satisfies $\lambda = \inf_k\{x_k/k\} \in [-\infty, \infty)$.

3.2 (continuation) Find reasonable conditions on the sequence (α_n) such that: the generalized inequality

$$x_{m+n} \leq x_m + x_n + \alpha_m, \qquad m, n \geq 1,$$

implies the existence of the limit $\lambda = \lim_{n\to\infty}\{x_n/n\}$.

3.3 [120] *Bond/site critical probabilities.* Let G be an infinite connected graph with maximal vertex degree Δ. Show that the critical probabilities for bond and site percolation on G satisfy

$$p_c^{\text{bond}} \leq p_c^{\text{site}} \leq 1 - (1 - p_c^{\text{bond}})^\Delta.$$

The second inequality is in fact valid with Δ replaced by $\Delta - 1$.

3.4 Show that bond percolation on a graph G may be reformulated in terms of site percolation on a graph derived suitably from G.

3.5 Show that the connective constant of $\mathbb{L}^2$ lies strictly between 2 and 3.

3.6 Show the strict inequality $p_c(d) < \vec{p}_c(d)$ for the critical probabilities of unoriented and oriented percolation on $\mathbb{L}^d$ with $d \geq 2$.

3.7 *One-dimensional percolation.* Each edge of the one-dimensional lattice $\mathbb{L}$ is declared *open* with probability p. For $k \in \mathbb{Z}$, let $r(k) = \max\{u : k \leftrightarrow k + u\}$, and $L_n = \max\{r(k) : 1 \leq k \leq n\}$. Show that $\mathbb{P}_p(L_n > u) \leq np^u$, and deduce that, for $\epsilon > 0$,

$$\mathbb{P}_p\left(L_n > \frac{(1+\epsilon)\log n}{\log(1/p)}\right) \to 0 \qquad \text{as } n \to \infty.$$

This is the famous problem of the longest run of heads in n tosses of a coin.

3.8 (continuation) Show that, for $\epsilon > 0$,

$$\mathbb{P}_p\left(L_n < \frac{(1-\epsilon)\log n}{\log(1/p)}\right) \to 0 \qquad \text{as } n \to \infty.$$

By suitable refinements of the error estimates above, show that, for $\epsilon > 0$,

$$\mathbb{P}_p\left(\frac{(1-\epsilon)\log n}{\log(1/p)} < L_n < \frac{(1+\epsilon)\log n}{\log(1/p)}, \text{ for all but finitely many } n\right) = 1.$$

3.9 [108] Each edge of the square lattice $\mathbb{L}^2$ is oriented in a random direction, horizontal edges being oriented eastwards with probability p and westwards otherwise, and vertical edges being oriented northwards with probability p and southwards otherwise. Let $\eta(p)$ be the probability that there exists an infinite oriented path starting at the origin. By coupling with undirected bond percolation, or otherwise, show that $\eta(\frac{1}{2}) = 0$.

It is an open problem to decide whether or not $\eta(p) > 0$ for $p \neq \frac{1}{2}$.

3.10 The vertex (i, j) of $\mathbb{L}^2$ is called *even* if $i + j$ is even, and *odd* otherwise. Vertical edges are oriented from the even endpoint to the odd, and horizontal edges vice versa. Each edge is declared *open* with probability p, and closed otherwise (independently between edges). Show that, for p sufficiently close to 1, there is strictly positive probability that the origin is the endpoint of an infinite open oriented path.

3.11 [111, 171, 172] A *word* is an element of the set $\{0, 1\}^{\mathbb{N}}$ of singly infinite 0/1 sequences. Let $p \in (0, 1)$ and $M \geq 1$. Consider oriented site percolation on $\mathbb{Z}^2$, in which the state $\omega(x)$ of a vertex x equals 1 with probability p, and 0 otherwise. A word $w = (w_1, w_2, \dots)$ is said to be *M-seen* if there exists an infinite oriented path $x_0 = 0, x_1, x_2, \dots$ of vertices such that $\omega(x_i) = w_i$ and $d(x_{i-1}, x_i) \leq M$ for $i \geq 1$. [Here, as usual, d denotes graph-theoretic distance.]

Calculate the probability that the square $\{1, 2, \dots, k\}^2$ contains both a 0 and a 1. Deduce by a block argument that

$$\psi_p(M) = \mathbb{P}_p(\text{all words are } M\text{-seen})$$

satisfies $\psi_p(M) > 0$ for $M \geq M(p)$, and determine an upper bound on the required $M(p)$.

4

Association and influence

Correlation inequalities have played a significant role in the theory of disordered spatial systems. The Holley inequality provides a sufficient condition for the stochastic ordering of two measures, and also a route to a proof of the famous FKG inequality. For product measures, the complementary BK inequality involves the concept of 'disjoint occurrence'. Two concepts of concentration are considered here. The Hoeffding inequality provides a bound on the tail of a martingale with bounded differences. Another concept of 'influence' proved by Kahn, Kalai, and Linial leads to sharp-threshold theorems for increasing events under either product or FKG measures.

4.1 Holley inequality

We review the stochastic ordering of probability measures on a discrete space. Let E be a non-empty finite set, and $\Omega = \{0, 1\}^E$. The sample space Ω is partially ordered by

$$\omega_1 \leq \omega_2 \quad \text{if} \quad \omega_1(e) \leq \omega_2(e) \text{ for all } e \in E.$$

A non-empty subset $A \subseteq \Omega$ is called *increasing* if

$$\omega \in A, \ \omega \leq \omega' \quad \Rightarrow \quad \omega' \in A,$$

and *decreasing* if

$$\omega \in A, \ \omega' \leq \omega \quad \Rightarrow \quad \omega' \in A.$$

If $A \ (\neq \Omega)$ is increasing, then its complement $\overline{A} = \Omega \setminus A$ is decreasing.

4.1 Definition. Given two probability measures μ_i, $i = 1, 2$, on Ω, we write $\mu_1 \leq_{\text{st}} \mu_2$ if

$$\mu_1(A) \leq \mu_2(A) \quad \text{for all increasing events } A.$$

Equivalently, $\mu_1 \leq_{\text{st}} \mu_2$ if and only if $\mu_1(f) \leq \mu_2(f)$ for all increasing functions $f : \Omega \to \mathbb{R}$. There is an important and useful result, often termed

Strassen's theorem, that asserts that measures satisfying $\mu_1 \leq_{st} \mu_2$ may be coupled in a 'pointwise monotone' manner. Such a statement is valid for very general spaces (see [173]), but we restrict ourselves here to the current context. The proof is omitted, and may be found in many places including [181, 237].

4.2 Theorem [227]. *Let μ_1 and μ_2 be probability measures on Ω. The following two statements are equivalent.*

(a) $\mu_1 \leq_{st} \mu_2$.

(b) *There exists a probability measure ν on Ω^2 such that*
$$\nu\big(\{(\pi, \omega) : \pi \leq \omega\}\big) = 1,$$
and whose marginal measures are μ_1 and μ_2.

For $\omega_1, \omega_2 \in \Omega$, we define the (pointwise) maximum and minimum configurations by

(4.3)
$$\omega_1 \vee \omega_2(e) = \max\{\omega_1(e), \omega_2(e)\},$$
$$\omega_1 \wedge \omega_2(e) = \min\{\omega_1(e), \omega_2(e)\},$$

for $e \in E$. A probability measure μ on Ω is called *positive* if $\mu(\omega) > 0$ for all $\omega \in \Omega$.

4.4 Theorem (Holley inequality) [140]. *Let μ_1 and μ_2 be positive probability measures on Ω satisfying*

(4.5) $\mu_2(\omega_1 \vee \omega_2)\mu_1(\omega_1 \wedge \omega_2) \geq \mu_1(\omega_1)\mu_2(\omega_2), \qquad \omega_1, \omega_2 \in \Omega.$

Then $\mu_1 \leq_{st} \mu_2$.

Condition (4.5) is not necessary for the stochastic inequality, but is equivalent to a stronger property of 'monotonicity', see [109, Thm 2.3].

Proof. The main step is the proof that μ_1 and μ_2 can be 'coupled' in such a way that the component with marginal measure μ_2 lies above (in the sense of sample realizations) that with marginal measure μ_1. This is achieved by constructing a certain Markov chain with the coupled measure as unique invariant measure.

Here is a preliminary calculation. Let μ be a positive probability measure on Ω. We can construct a time-reversible Markov chain with state space Ω and unique invariant measure μ by choosing a suitable generator G satisfying the detailed balance equations. The dynamics of the chain involve the 'switching on or off' of single components of the current state.

For $\omega \in \Omega$ and $e \in E$, we define the configurations ω^e, ω_e by

(4.6) $\omega^e(f) = \begin{cases} \omega(f) & \text{if } f \neq e, \\ 1 & \text{if } f = e, \end{cases} \qquad \omega_e(f) = \begin{cases} \omega(f) & \text{if } f \neq e, \\ 0 & \text{if } f = e. \end{cases}$

Let $G : \Omega^2 \to \mathbb{R}$ be given by

$$(4.7) \qquad G(\omega_e, \omega^e) = 1, \quad G(\omega^e, \omega_e) = \frac{\mu(\omega_e)}{\mu(\omega^e)},$$

for all $\omega \in \Omega$, $e \in E$. Set $G(\omega, \omega') = 0$ for all other pairs ω, ω' with $\omega \neq \omega'$. The diagonal elements are chosen in such a way that

$$\sum_{\omega' \in \Omega} G(\omega, \omega') = 0, \qquad \omega \in \Omega.$$

It is elementary that

$$\mu(\omega)G(\omega, \omega') = \mu(\omega')G(\omega', \omega), \qquad \omega, \omega' \in \Omega,$$

and therefore G generates a time-reversible Markov chain on the state space Ω. This chain is irreducible (using (4.7)), and therefore possesses a unique invariant measure μ (see [121, Thm 6.5.4]).

We next follow a similar route for *pairs* of configurations. Let μ_1 and μ_2 satisfy the hypotheses of the theorem, and let S be the set of all pairs (π, ω) of configurations in Ω satisfying $\pi \leq \omega$. We define $H : S \times S \to \mathbb{R}$ by

$$(4.8) \qquad H(\pi_e, \omega; \pi^e, \omega^e) = 1,$$

$$(4.9) \qquad H(\pi, \omega^e; \pi_e, \omega_e) = \frac{\mu_2(\omega_e)}{\mu_2(\omega^e)},$$

$$(4.10) \qquad H(\pi^e, \omega^e; \pi_e, \omega^e) = \frac{\mu_1(\pi_e)}{\mu_1(\pi^e)} - \frac{\mu_2(\omega_e)}{\mu_2(\omega^e)},$$

for all $(\pi, \omega) \in S$ and $e \in E$; all other off-diagonal values of H are set to 0. The diagonal terms are chosen in such a way that

$$\sum_{\pi', \omega'} H(\pi, \omega; \pi', \omega') = 0, \qquad (\pi, \omega) \in S.$$

Equation (4.8) specifies that, for $\pi \in \Omega$ and $e \in E$, the edge e is acquired by π (if it does not already contain it) at rate 1; any edge so acquired is added also to ω if it does not already contain it. (Here, we speak of a configuration ψ containing an edge e if $\psi(e) = 1$.) Equation (4.9) specifies that, for $\omega \in \Omega$ and $e \in E$ with $\omega(e) = 1$, the edge e is removed from ω (and also from π if $\pi(e) = 1$) at the rate given in (4.9). For e with $\pi(e) = 1$, there is an additional rate given in (4.10) at which e is removed from π but not from ω. We need to check that this additional rate is indeed non-negative, and the required inequality,

$$\mu_2(\omega^e)\mu_1(\pi_e) \geq \mu_1(\pi^e)\mu_2(\omega_e), \qquad \pi \leq \omega,$$

follows from (and is indeed equivalent to) assumption (4.5).

Let $(X_t, Y_t)_{t \geq 0}$ be a Markov chain on S with generator H, and set $(X_0, Y_0) = (0, 1)$, where 0 (respectively, 1) is the state of all 0s (respectively, 1s). By examination of (4.8)–(4.10), we see that $X = (X_t)_{t \geq 0}$ is a Markov chain with generator given by (4.7) with $\mu = \mu_1$, and that $Y = (Y_t)_{t \geq 0}$ arises similarly with $\mu = \mu_2$.

Let κ be an invariant measure for the paired chain $(X_t, Y_t)_{t \geq 0}$. Since X and Y have (respective) unique invariant measures μ_1 and μ_2, the marginals of κ are μ_1 and μ_2. We have by construction that $\kappa(S) = 1$, and κ is the required 'coupling' of μ_1 and μ_2.

Let $(\pi, \omega) \in S$ be chosen according to the measure κ. Then

$$\mu_1(f) = \kappa(f(\omega)) \leq \kappa(f(\pi)) = \mu_2(f),$$

for any increasing function f. Therefore, $\mu_1 \leq_{st} \mu_2$. $\qquad\square$

4.2 FKG inequality

The FKG inequality for product measures was discovered by Harris [135], and is often named now after the authors Fortuin, Kasteleyn, and Ginibre of [91] who proved the more general version that is the subject of this section. See the appendix of [109] for a historical account. Let E be a finite set, and $\Omega = \{0, 1\}^E$ as usual.

4.11 Theorem (FKG inequality) [91]. *Let μ be a positive probability measure on Ω such that*

$$(4.12) \qquad \mu(\omega_1 \vee \omega_2)\mu(\omega_1 \wedge \omega_2) \geq \mu(\omega_1)\mu(\omega_2), \qquad \omega_1, \omega_2 \in \Omega.$$

Then μ is 'positively associated' in that

$$(4.13) \qquad\qquad \mu(fg) \geq \mu(f)\mu(g)$$

for all increasing random variables $f, g : \Omega \to \mathbb{R}$.

It is explained in [91] how the condition of (strict) positivity can be removed. Condition (4.12) is sometimes called the 'FKG lattice condition'.

Proof. Assume that μ satisfies (4.12), and let f and g be increasing functions. By adding a constant to the function g, we see that it suffices to prove (4.13) under the additional hypothesis that g is strictly positive. Assume the last holds. Define positive probability measures μ_1 and μ_2 on Ω by $\mu_1 = \mu$ and

$$\mu_2(\omega) = \frac{g(\omega)\mu(\omega)}{\sum_{\omega'} g(\omega')\mu(\omega')}, \qquad \omega \in \Omega.$$

Since g is increasing, the Holley condition (4.5) follows from (4.12). By the Holley inequality, Theorem 4.4,

$$\mu_1(f) \le \mu_2(f),$$

which is to say that

$$\frac{\sum_\omega f(\omega)g(\omega)\mu(\omega)}{\sum_{\omega'} g(\omega')\mu(\omega')} \ge \sum_\omega f(\omega)\mu(\omega)$$

as required. □

4.3 BK inequality

In the special case of product measure on Ω, there is a type of converse inequality to the FKG inequality, named the BK inequality after van den Berg and Kesten [35]. This is based on a concept of 'disjoint occurrence' that we make more precise as follows.

For $\omega \in \Omega$ and $F \subseteq E$, we define the cylinder event $C(\omega, F)$ generated by ω on F by

$$C(\omega, F) = \{\omega' \in \Omega : \omega'(e) = \omega(e) \text{ for all } e \in F\}$$

$$= (\omega(e) : e \in F) \times \{0, 1\}^{E \setminus F}.$$

We define the event $A \square B$ as the set of all $\omega \in \Omega$ for which there exists a set $F \subseteq E$ such that $C(\omega, F) \subseteq A$ and $C(\omega, E \setminus F) \subseteq B$. Thus, $A \square B$ is the set of configurations ω for which there exist disjoint sets F, G of indices with the property that: knowledge of ω restricted to F (respectively, G) implies that $\omega \in A$ (respectively, $\omega \in B$). In the special case when A and B are increasing, $C(\omega, F) \subseteq A$ if and only if $\omega_F \in A$, where

$$\omega_F(e) = \begin{cases} \omega(e) & \text{for } e \in F, \\ 0 & \text{for } e \notin F. \end{cases}$$

Thus, in this case, $A \square B = A \circ B$, where

$$A \circ B = \{\omega : \text{there exists } F \subseteq E \text{ such that } \omega_F \in A, \ \omega_{E \setminus F} \in B\}.$$

The set F is permitted to depend on the choice of configuration ω.

Three notes about disjoint occurrence:

(4.14) $A \square B \subseteq A \cap B$,

(4.15)   if A and B are increasing, then so is $A \square B (= A \circ B)$,

(4.16)   if A increasing and B decreasing, then $A \square B = A \cap B$.

Let $\mathbb{P}$ be the product measure on Ω with local densities $p_e, e \in E$, that is

$$\mathbb{P} = \prod_{e \in E} \mu_e,$$

where $\mu_e(0) = 1 - p_e$ and $\mu_e(1) = p_e$.

4.17 Theorem (BK inequality) [35]. *For increasing subsets A, B of Ω,*

(4.18) $$\mathbb{P}(A \circ B) \le \mathbb{P}(A)\mathbb{P}(B).$$

It is not known for which non-product measures (4.18) holds. It seems reasonable, for example, to conjecture that (4.18) holds for the measure $\mathbb{P}_k$ that selects a k-subset of E uniformly at random. It would be very useful to show that the random-cluster measure $\phi_{p,q}$ on Ω satisfies (4.18) whenever $0 < q < 1$, although we may have to survive with rather less. See Chapter 8, and [109, Sect. 3.9].

The conclusion of the BK inequality is in fact valid for all pairs A, B of events, regardless of whether or not they are increasing. This is much harder to prove, and has not yet been as valuable as originally expected in the analysis of disordered systems.

4.19 Theorem (Reimer inequality) [206]. *For A, B $\subseteq \Omega$,*

$$\mathbb{P}(A \,\square\, B) \le \mathbb{P}(A)\mathbb{P}(B).$$

Let A and B be increasing. By applying Reimer's inequality to the events A and $\overline{B}$, we obtain by (4.16) that $\mathbb{P}(A \cap B) \ge \mathbb{P}(A)\mathbb{P}(B)$. Therefore, Reimer's inequality includes both the FKG and BK inequalities for the product measure $\mathbb{P}$. The proof of Reimer's inequality is omitted, see [50, 206].

Proof of Theorem 4.17. We present the 'simple' proof of [33, 106, 237]. Those who prefer proofs by induction are directed to [48]. Let $1, 2, \dots, N$ be an ordering of E. We shall consider the duplicated sample space $\Omega \times \Omega'$, where $\Omega = \Omega' = \{0, 1\}^E$, with which we associate the product measure $\widehat{\mathbb{P}} = \mathbb{P} \times \mathbb{P}$. Elements of Ω (respectively, Ω') are written as ω (respectively, ω'). Let A and B be increasing subsets of $\{0, 1\}^E$. For $1 \le j \le N + 1$ and $(\omega, \omega') \in \Omega \times \Omega'$, define the N-vector ω_j by

$$\omega_j = \big(\omega'(1), \omega'(2), \dots, \omega'(j-1), \omega(j), \dots, \omega(N)\big),$$

so that the ω_j interpolate between $\omega_1 = \omega$ and $\omega_{N+1} = \omega'$. Let the events $\widehat{A}_j$, $\widehat{B}$ of $\Omega \times \Omega'$ be given by

$$\widehat{A}_j = \{(\omega, \omega') : \omega_j \in A\}, \qquad \widehat{B} = \{(\omega, \omega') : \omega \in B\}.$$

Note that:

(a) $\widehat{A}_1 = A \times \Omega'$ and $\widehat{B} = B \times \Omega'$, so that $\widehat{\mathbb{P}}(\widehat{A}_1 \circ \widehat{B}) = \mathbb{P}(A \circ B)$,

(b) $\widehat{A}_{N+1}$ and $\widehat{B}$ are defined in terms of disjoint subsets of E, so that

$$\widehat{\mathbb{P}}(\widehat{A}_{N+1} \circ \widehat{B}) = \widehat{\mathbb{P}}(\widehat{A}_{N+1})\widehat{\mathbb{P}}(\widehat{B}) = \mathbb{P}(A)\mathbb{P}(B).$$

It thus suffices to show that

(4.20) $\widehat{\mathbb{P}}(\widehat{A}_j \circ \widehat{B}) \leq \widehat{\mathbb{P}}(\widehat{A}_{j+1} \circ \widehat{B}),$ $1 \leq j \leq N,$

and this we do, for given j, by conditioning on the values of the $\omega(i)$, $\omega'(i)$ for all $i \neq j$. Suppose these values are given, and classify them as follows. There are three cases.

1. $\widehat{A}_j \circ \widehat{B}$ does not occur when $\omega(j) = \omega'(j) = 1$.
2. $\widehat{A}_j \circ \widehat{B}$ occurs when $\omega(j) = \omega'(j) = 0$, in which case $\widehat{A}_{j+1} \circ \widehat{B}$ occurs also.
3. Neither of the two cases above hold.

Consider the third case. Since $\widehat{A}_j \circ \widehat{B}$ does not depend on the value $\omega'(j)$, we have in this case that $\widehat{A}_j \circ B$ occurs if and only if $\omega(j) = 1$, and therefore the conditional probability of $\widehat{A}_j \circ \widehat{B}$ is p_j. When $\omega(j) = 1$, edge j is 'contributing' to either $\widehat{A}_j$ or $\widehat{B}$ but not both. Replacing $\omega(j)$ by $\omega'(j)$, we find similarly that the conditional probability of $\widehat{A}_{j+1} \circ \widehat{B}$ is at least p_j.

In each of the three cases above, the conditional probability of $\widehat{A}_j \circ \widehat{B}$ is no greater than that of $\widehat{A}_{j+1} \circ \widehat{B}$, and (4.20) follows.   □

4.4 Hoeffding inequality

Let $(Y_n, \mathcal{F}_n)$, $n \geq 0$, be a martingale. We can obtain bounds for the tail of Y_n in terms of the sizes of the martingale differences $D_k = Y_k - Y_{k-1}$. These bounds are surprisingly tight, and they have had substantial impact in various areas of application, especially those with a combinatorial structure. We describe such a bound in this section for the case when the D_k are bounded random variables.

4.21 Theorem (Hoeffding inequality). *Let $(Y_n, \mathcal{F}_n)$, $n \geq 0$, be a martingale such that $|Y_k - Y_{k-1}| \leq K_k$ (a.s.) for all k and some real sequence (K_k). Then*

$$\mathbb{P}(Y_n - Y_0 \geq x) \leq \exp\left(-\tfrac{1}{2}x^2/L_n\right), \qquad x > 0,$$

where $L_n = \sum_{k=1}^{n} K_k^2$.

Since Y_n is a martingale, so is $-Y_n$, and thus the same bound is valid for $\mathbb{P}(Y_n - Y_0 \leq -x)$. Such inequalities are often named after Azuma [22] and Hoeffding [139].

Theorem 4.21 is one of a family of inequalities frequently used in probabilistic combinatorics, in what is termed the 'method of bounded differences'. See the discussion in [186]. Its applications are of the following general form. Suppose that we are given N random variables

$X_1, X_2, \ldots, X_N$, and we wish to study the behaviour of some function $Z = Z(X_1, X_2, \ldots, X_N)$. For example, the X_i might be the sizes of objects to be packed into bins, and Z the minimum number of bins required to pack them. Let $\mathcal{F}_n = \sigma(X_1, X_2, \ldots, X_n)$, and define the martingale $Y_n = \mathbb{E}(Z \mid \mathcal{F}_n)$. Thus, $Y_0 = \mathbb{E}(Z)$ and $Y_N = Z$. If the martingale differences are bounded, Theorem 4.21 provides a bound for the tail probability $\mathbb{P}(|Z - \mathbb{E}(Z)| \geq x)$. We shall see an application of this type at Theorem 11.13, which deals with the chromatic number of random graphs. Further applications may be found in [121, Sect. 12.2], for example.

Proof. The function $g(d) = e^{\psi d}$ is convex for $\psi > 0$, and therefore

(4.22) $\qquad e^{\psi d} \leq \frac{1}{2}(1 - d)e^{-\psi} + \frac{1}{2}(1 + d)e^{\psi}, \qquad |d| \leq 1.$

Applying this to a random variable D having mean 0 and satisfying $\mathbb{P}(|D| \leq 1) = 1$, we obtain

(4.23) $\qquad \mathbb{E}(e^{\psi D}) \leq \frac{1}{2}(e^{-\psi} + e^{\psi}) < e^{\frac{1}{2}\psi^2}, \qquad \psi > 0,$

where the final inequality is shown by a comparison of the coefficients of the powers ψ^{2n}.

By Markov's inequality,

(4.24) $\qquad \mathbb{P}(Y_n - Y_0 \geq x) \leq e^{-\theta x}\mathbb{E}(e^{\theta(Y_n - Y_0)}), \qquad \theta > 0.$

With $D_n = Y_n - Y_{n-1}$,

$$\mathbb{E}(e^{\theta(Y_n - Y_0)}) = \mathbb{E}(e^{\theta(Y_{n-1} - Y_0)}e^{\theta D_n}).$$

Since $Y_{n-1} - Y_0$ is $\mathcal{F}_{n-1}$-measurable,

(4.25) $\qquad \mathbb{E}(e^{\theta(Y_n - Y_0)} \mid \mathcal{F}_{n-1}) = e^{\theta(Y_{n-1} - Y_0)}\mathbb{E}(e^{\theta D_n} \mid \mathcal{F}_{n-1})$

$$\leq e^{\theta(Y_{n-1} - Y_0)} \exp\left(\tfrac{1}{2}\theta^2 K_n^2\right),$$

by (4.23) applied to the random variable D_n/K_n. Take expectations of (4.25) and iterate to obtain

$$\mathbb{E}(e^{\theta(Y_n - Y_0)}) \leq \mathbb{E}(e^{\theta(Y_{n-1} - Y_0)}) \exp\left(\tfrac{1}{2}\theta^2 K_n^2\right) \leq \exp\left(\tfrac{1}{2}\theta^2 L_n\right).$$

Therefore, by (4.24),

$$\mathbb{P}(Y_n - Y_0 \geq x) \leq \exp\left(-\theta x + \tfrac{1}{2}\theta^2 L_n\right), \qquad \theta > 0.$$

Let $x > 0$, and set $\theta = x/L_n$ (this is the value that minimizes the exponent). Then

$$\mathbb{P}(Y_n - Y_0 \geq x) \leq \exp\left(-\tfrac{1}{2}x^2/L_n\right), \qquad x > 0,$$

as required. $\qquad \square$

4.5 Influence for product measures

Let $N \geq 1$ and $E = \{1, 2, \dots, N\}$, and write $\Omega = \{0, 1\}^E$. Let μ be a probability measure on Ω, and A an event (that is, a subset of Ω). Two ways of defining the 'influence' of an element $e \in E$ on the event A come to mind. The (*conditional*) *influence* is defined to be

$$(4.26) \qquad J_A(e) = \mu(A \mid \omega(e) = 1) - \mu(A \mid \omega(e) = 0).$$

The *absolute influence* is

$$(4.27) \qquad I_A(e) = \mu(1_A(\omega^e) \neq 1_A(\omega_e)),$$

where 1_A is the indicator function of A, and ω^e, ω_e are the configurations given by (4.6). In a voting analogy, each of N voters has 1 vote, and A is the set of vote-vectors that result in a given outcome. Then $I_A(e)$ is the probability that voter e can influence the outcome.

We make two remarks concerning the above definitions. First, if A is increasing,

$$(4.28) \qquad I_A(e) = \mu(A^e) - \mu(A_e),$$

where

$$A^e = \{\omega \in \Omega : \omega^e \in A\}, \qquad A_e = \{\omega \in \Omega : \omega_e \in A\}.$$

If, in addition, μ is a product measure, then $I_A(e) = J_A(e)$. Note that influences depend on the underlying measure.

Let ϕ_p be product measure with density p on Ω, and write $\phi = \phi_{\frac{1}{2}}$, the uniform measure. All logarithms are taken to base 2 until further notice.

There has been extensive study of the largest (absolute) influence, namely $\max_e I_A(e)$, when μ is a product measure, and this has been used to obtain 'sharp threshold' theorems for the probability $\phi_p(A)$ of an increasing event A viewed as a function of p. The principal theorems are given in this section, with proofs in the next. The account presented here differs in a number of respects from the original references.

4.29 Theorem (Influence) [145]. *There exists a constant $c \in (0, \infty)$ such that the following holds. Let $N \geq 1$, let E be a finite set with $|E| = N$, and let A be a subset of $\Omega = \{0, 1\}^E$ with $\phi(A) \in (0, 1)$. Then*

$$(4.30) \qquad \sum_{e \in E} I_A(e) \geq c\phi(A)(1 - \phi(A)) \log[1/\max_e I_A(e)],$$

where the reference measure is $\phi = \phi_{\frac{1}{2}}$. There exists $e \in E$ such that

$$(4.31) \qquad I_A(e) \geq c\phi(A)(1 - \phi(A))\frac{\log N}{N}.$$

Note that

$$\phi(A)(1 - \phi(A)) \geq \tfrac{1}{2} \min\{\phi(A), 1 - \phi(A)\}.$$

We indicate at this stage the reason why (4.30) implies (4.31). We may assume that $m = \max_e I_A(e)$ satisfies $m > 0$, since otherwise

$$\phi(A)(1 - \phi(A)) = 0.$$

Since

$$\sum_{e \in E} I_A(e) \leq Nm,$$

we have by (4.30) that

$$\frac{m}{\log(1/m)} \geq \frac{c\phi(A)(1 - \phi(A))}{N}.$$

Inequality (4.31) follows with an amended value of c, by the monotonicity of $m / \log(1/m)$ or otherwise.[1]

Such results have applications to several topics including random graphs, random walks, and percolation, see [147]. We summarize two such applications next, and we defer until Section 5.8 an application to site percolation on the triangular lattice.

I. *First-passage percolation.* This is the theory of passage times on a graph whose edges have random 'travel-times'. Suppose we assign to each edge e of the d-dimensional cubic lattice $\mathbb{L}^d$ a random travel-time T_e, the T_e being non-negative and independent with common distribution function F. The passage time of a path π is the sum of the travel-times of its edges. Given two vertices u, v, the passage time $T_{u,v}$ is defined as the infimum of the passage times of the set of paths joining u to v. The main question is to understand the asymptotic properties of $T_{0,v}$ as $|v| \to \infty$. This model for the time-dependent flow of material was introduced in [131], and has been studied extensively since.

It is a consequence of the subadditive ergodic theorem that, subject to a suitable moment condition, the (deterministic) limit

$$\mu_v = \lim_{n \to \infty} \frac{1}{n} T_{0,nv}$$

exists almost surely. Indeed, the subadditive ergodic theorem was conceived explicitly in order to prove such a statement for first-passage percolation. The constant μ_v is called the *time constant* in direction v. One of the open problems is to understand the asymptotic behaviour of $\mathrm{var}(T_{0,v})$ as

---

[1] When $N = 1$, there is nothing to prove. This is left as an exercise when $N \geq 2$.

$|v| \to \infty$. Various relevant results are known, and one of the best uses an influence theorem due to Talagrand [231] and related to Theorem 4.29. Specifically, it is proved in [31] that $\text{var}(T_{0,v}) \leq C|v|/\log|v|$ for some constant $C = C(a, b, d)$, in the situation when each T_e is equally likely to take either of the two positive values a, b. It has been predicted that $\text{var}(T_{0,v}) \sim |v|^{2/3}$ when $d = 2$. This work has been continued in [29].

II. *Voronoi percolation model.* This continuum model is constructed as follows in $\mathbb{R}^2$. Let Π be a Poisson process of intensity 1 in $\mathbb{R}^2$. With any $u \in \Pi$, we associate the 'tile'

$$T_u = \{x \in \mathbb{R}^2 : |x - u| \leq |x - v| \text{ for all } v \in \Pi\}.$$

Two points $u, v \in \Pi$ are declared *adjacent*, written $u \sim v$, if T_u and T_v share a boundary segment. We now consider site percolation on the graph Π with this adjacency relation. It was long believed that the critical percolation probability of this model is $\frac{1}{2}$ (almost surely, with respect to the Poisson measure), and this was proved by Bollobás and Riordan [46] using a version of the threshold Theorem 4.82 that is consequent on Theorem 4.29.

Bollobás and Riordan showed also in [47] that a similar argument leads to an approach to the proof that the critical probability of bond percolation on $\mathbb{Z}^2$ equals $\frac{1}{2}$. They used Theorem 4.82 in place of Kesten's explicit proof of sharp threshold for this model, see [151, 152]. A "shorter" version of [47] is presented in Section 5.8 for the case of site percolation on the triangular lattice.

We return to the influence theorem and its ramifications. There are several useful references concerning influence for product measures, see [92, 93, 145, 147, 150] and their bibliographies.[2] The order of magnitude $N^{-1} \log N$ is the best possible in (4.31), as shown by the following 'tribes' example taken from [30]. A population of N individuals comprises t 'tribes' each of cardinality $s = \log N - \log \log N + \alpha$. Each individual votes 1 with probability $\frac{1}{2}$ and otherwise 0, and different individuals vote independently of one another. Let A be the event that there exists a tribe all of whose members vote 1. It is easily seen that

$$1 - P(A) = \left(1 - \frac{1}{2^s}\right)^t$$
$$\sim e^{-t/2^s} \sim e^{-1/2^\alpha},$$

---

[2]The treatment presented here makes heavy use of the work of the 'Israeli' school. The earlier paper of Russo [213] must not be overlooked, and there are several important papers of Talagrand [230, 231, 232, 233]. Later approaches to Theorem 4.29 can be found in [84, 208, 209].

and, for all i,

$$I_A(i) = \left(1 - \frac{1}{2^s}\right)^{t-1} \frac{1}{2^{s-1}}$$

$$\sim e^{-1/2^\alpha} 2^{-\alpha-1} \frac{\log N}{N},$$

The 'basic' Theorem 4.29 on the discrete cube $\Omega = \{0, 1\}^E$ can be extended to the 'continuum' cube $K = [0, 1]^E$, and hence to other product spaces. We state the result for K next. Let λ be uniform (Lebesgue) measure on K. For a measurable subset $A \subseteq K$, it is usual (see, for example, [51]) to define the influence of $e \in E$ on A as

$$L_A(e) = \lambda_{N-1}\big(\{\omega \in K : 1_A(\omega) \text{ is a non-constant function of } \omega(e)\}\big).$$

That is, $L_A(e)$ is the $(N-1)$-dimensional Lebesgue measure of the set of all $\psi \in [0, 1]^{E \setminus \{e\}}$ with the property that: both A and its complement $\overline{A}$ intersect the 'fibre'

$$F_\psi = \{\psi\} \times [0, 1] = \{\omega \in K : \omega(f) = \psi(f), \ f \neq e\}.$$

It is more natural to consider elements ψ for which $A \cap F_\psi$ has Lebesgue measure strictly between 0 and 1, and thus we define the influence in these notes by

(4.32) $I_A(e) = \lambda_{N-1}\big(\{\psi \in [0, 1]^{E \setminus \{e\}} : 0 < \lambda_1(A \cap F_\psi) < 1\}\big).$

Here and later, when convenient, we write λ_k for k-dimensional Lebesgue measure. Note that $I_A(e) \leq L_A(e)$.

4.33 Theorem [51]. *There exists a constant $c \in (0, \infty)$ such that the following holds. Let $N \geq 1$, let E be a finite set with $|E| = N$, and let A be an increasing subset of the cube $K = [0, 1]^E$ with $\lambda(A) \in (0, 1)$. Then*

(4.34) $$\sum_{e \in E} I_A(e) \geq c\lambda(A)(1 - \lambda(A)) \log[1/(2m)],$$

where $m = \max_e I_A(e)$, and the reference measure on K is Lebesgue measure λ. There exists $e \in E$ such that

(4.35) $$I_A(e) \geq c\lambda(A)(1 - \lambda(A)) \frac{\log N}{N}.$$

We shall see in Theorem 4.38 that the condition of monotonicity of A can be removed. The factor '2' in (4.34) is innocent in the following regard. The inequality is important only when m is small, and, for $m \leq \frac{1}{3}$ say, we may remove the '2' and replace c by a larger constant.

Results similar to those of Theorems 4.29 and 4.33 have been proved in [100] for certain non-product measures, and all increasing events. Let μ be a positive probability measure on the discrete space $\Omega = \{0, 1\}^E$ satisfying the FKG lattice condition (4.12). For any increasing subset A of Ω with $\mu(A) \in (0, 1)$, we have that

$$(4.36) \qquad \sum_{e \in E} J_A(e) \geq c\mu(A)(1 - \mu(A)) \log[1/(2m)],$$

where $m = \max_e J_A(e)$. Furthermore, as above, there exists $e \in E$ such that

$$(4.37) \qquad J_A(e) \geq c\mu(A)(1 - \mu(A)) \frac{\log N}{N}.$$

Note the use of *conditional* influence $J_A(e)$, with non-product reference measure μ. Indeed, (4.37) can fail for all e when J_A is replaced by I_A. The proof of (4.36) makes use of Theorem 4.33, and is omitted here, see [100, 101].

The domain of Theorem 4.33 can be extended to powers of an arbitrary probability space, that is with $([0, 1], \lambda_1)$ replaced by a general probability space. Let $|E| = N$ and let $X = (\Sigma, \mathcal{F}, P)$ be a probability space. We write X^E for the product space of X. Let $A \subseteq \Sigma^E$ be measurable. The influence of $e \in E$ is given as in (4.32) by

$$I_A(e) = \mathbb{P}\big(\{\psi \in \Sigma^{E \setminus \{e\}} : 0 < P(A \cap F_\psi) < 1\}\big),$$

with $\mathbb{P} = P^E$ and $F_\psi = \{\psi\} \times \Sigma$, the 'fibre' of all $\omega \in X^E$ such that $\omega(f) = \psi(f)$ for $f \neq e$.

The following theorem contains two statements: that the influence inequalities are valid for general product spaces, and that they hold for non-increasing events. We shall require a condition on $X = (\Sigma, \mathcal{F}, P)$ for the first of these, and we state this next. The pair $(\mathcal{F}, P)$ generates a measure ring (see [126, §40] for the relevant definitions). We call this measure ring *separable* if it is separable when viewed as a metric space with metric $\rho(B, B') = P(B \triangle B')$.[3]

4.38 Theorem [51]. *Let $X = (\Sigma, \mathcal{F}, P)$ be a probability space whose non-atomic part has a separable measure ring. Let $N \geq 1$, let E be a finite set with $|E| = N$, and let $A \subseteq \Sigma^E$ be measurable in the product space X^E, with $\mathbb{P}(A) \in (0, 1)$. There exists an absolute constant $c \in (0, \infty)$ such that*

$$(4.39) \qquad \sum_{e \in E} I_A(e) \geq c\mathbb{P}(A)(1 - \mathbb{P}(A)) \log[1/(2m)],$$

---

[3]A metric space is called *separable* if it possesses a countable dense subset. The condition of separability of Theorem 4.38 is omitted from [51].

where $m = \max_e I_A(e)$, and the reference measure is $\mathbb{P} = P^E$. There exists $e \in E$ with

(4.40) $$I_A(e) \geq c\mathbb{P}(A)(1 - \mathbb{P}(A))\frac{\log N}{N}.$$

Of especial interest is the case when $\Sigma = \{0, 1\}$ and P is Bernoulli measure with density p. Note that the atomic part of X is always separable, since there can be at most countably many atoms.

4.6 Proofs of influence theorems

This section contains the proofs of the theorems of the last.

Proof of Theorem 4.29. We use a (discrete) Fourier analysis of functions $f : \Omega \to \mathbb{R}$. Define the inner product by

$$\langle f, g \rangle = \phi(fg), \qquad f, g : \Omega \to \mathbb{R},$$

where $\phi = \phi_{\frac{1}{2}}$, so that the L^2-norm of f is given by

$$\|f\|_2 = \sqrt{\phi(f^2)} = \sqrt{\langle f, f \rangle}.$$

We call f *Boolean* if it takes values in the set $\{0, 1\}$. Boolean functions are in one–one correspondence with the power set of E via the relation $f = 1_A \leftrightarrow A$. If f is Boolean, say $f = 1_A$, then

(4.41) $$\|f\|_2^2 = \phi(f^2) = \phi(f) = \phi(A).$$

For $F \subseteq E$, let

$$u_F(\omega) = \prod_{e \in F}(-1)^{\omega(e)} = (-1)^{\sum_{e \in F} \omega(e)}, \qquad \omega \in \Omega.$$

It can be checked that the functions u_F, $F \subseteq E$, form an orthonormal basis for the function space. Thus, a function $f : \Omega \to \mathbb{R}$ may be expressed in the form

$$f = \sum_{F \subseteq E} \hat{f}(F)u_F,$$

where the so-called Fourier–Walsh coefficients of f are given by

$$\hat{f}(F) = \langle f, u_F \rangle, \qquad F \subseteq E.$$

In particular,

$$\hat{f}(\varnothing) = \phi(f),$$

and

$$\langle f, g \rangle = \sum_{F \subseteq E} \hat{f}(F)\hat{g}(F),$$

and the latter yields the Parseval relation

$$\|f\|_2^2 = \sum_{F \subseteq E} \hat{f}(F)^2. \tag{4.42}$$

Fourier analysis operates harmoniously with influences as follows. For $f = 1_A$ and $e \in E$, let

$$f_e(\omega) = f(\omega) - f(\kappa_e \omega),$$

where $\kappa_e \omega$ is the configuration ω with the state of e flipped. Since f_e takes values in the set $\{-1, 0, +1\}$, we have that $|f_e| = f_e^2$. The Fourier–Walsh coefficients of f_e are given by

$$
\begin{aligned}
\hat{f}_e(F) = \langle f_e, u_F \rangle &= \sum_{\omega \in \Omega} \frac{1}{2^N} [f(\omega) - f(\kappa_e \omega)](-1)^{|B \cap F|} \\
&= \sum_{\omega \in \Omega} \frac{1}{2^N} f(\omega) \big[(-1)^{|B \cap F|} - (-1)^{|(B \triangle \{e\}) \cap F|} \big],
\end{aligned}
$$

where $B = \eta(\omega) := \{e \in E : \omega(e) = 1\}$ is the set of ω-open indices. Now,

$$
\big[(-1)^{|B \cap F|} - (-1)^{|(B \triangle \{e\}) \cap F|} \big] =
\begin{cases}
0 & \text{if } e \notin F, \\
2(-1)^{|B \cap F|} = 2 u_F(\omega) & \text{if } e \in F,
\end{cases}
$$

so that

$$
\hat{f}_e(F) =
\begin{cases}
0 & \text{if } e \notin F, \\
2 \hat{f}(F) & \text{if } e \in F.
\end{cases}
\tag{4.43}
$$

The influence $I(e) = I_A(e)$ is the mean of $|f_e| = f_e^2$, so that, by (4.42),

$$I(e) = \|f_e\|_2^2 = 4 \sum_{F : e \in F} \hat{f}(F)^2, \tag{4.44}$$

and the total influence is

$$\sum_{e \in E} I(e) = 4 \sum_{F \subseteq E} |F| \hat{f}(F)^2. \tag{4.45}$$

We propose to find an upper bound for the sum $\phi(A) = \sum_F \hat{f}(F)^2$. From (4.45), we will extract an upper bound for the contributions to this sum from the $\hat{f}(F)^2$ for large $|F|$. This will be combined with a corresponding estimate for small $|F|$ that will be obtained as follows by considering a re-weighted sum $\sum_F \hat{f}(F)^2 \rho^{2|F|}$ for $0 < \rho < 1$.

For $w \in [1, \infty)$, we define the L^w-norm

$$\|g\|_w = \phi(|g|^w)^{1/w}, \qquad g : \Omega \to \mathbb{R},$$

recalling that $\|g\|_w$ is non-decreasing in w. For $\rho \in \mathbb{R}$, let $T_\rho g$ be the function

$$T_\rho g = \sum_{F \subseteq E} \hat{g}(F) \rho^{|F|} u_F,$$

so that

$$\|T_\rho g\|_2^2 = \sum_{F \subseteq E} \hat{g}(F)^2 \rho^{2|F|}.$$

When $\rho \in [-1, 1]$, $T_\rho g$ has a probabilistic interpretation. For $\omega \in \Omega$, let $\Psi_\omega = (\Psi_\omega(e) : e \in E)$ be a vector of independent random variables with

$$\Psi_\omega(e) = \begin{cases} \omega(e) & \text{with probability } \frac{1}{2}(1 + \rho), \\ 1 - \omega(e) & \text{otherwise.} \end{cases}$$

We claim that

(4.46) $$T_\rho g(\omega) = \mathbb{E}(g(\Psi_\omega)),$$

thus explaining why T_ρ is sometimes called the 'noise operator'. Equation (4.46) is proved as follows. First, for $F \subseteq E$,

$$\begin{aligned} \mathbb{E}(u_F(\Psi_\omega)) &= \mathbb{E}\left(\prod_{e \in F}(-1)^{\Psi_\omega(e)}\right) \\ &= \prod_{e \in F}(-1)^{\omega(e)}\left[\tfrac{1}{2}(1 + \rho) - \tfrac{1}{2}(1 - \rho)\right] \\ &= \rho^{|F|} u_F(\omega). \end{aligned}$$

Now, $g = \sum_F \hat{g}(F) u_F$, so that

$$\begin{aligned} \mathbb{E}(g(\Psi_\omega)) &= \sum_{F \subseteq E} \hat{g}(F)\mathbb{E}(u_F(\Psi_\omega)) \\ &= \sum_{F \subseteq E} \hat{g}(F)\rho^{|F|} u_F(\omega) = T_\rho g(\omega), \end{aligned}$$

as claimed at (4.46).

The next proposition is pivotal for the proof of the theorem. It is sometimes referred to as the 'hypercontractivity' lemma, and it is related to the log-Sobolev inequality. It is commonly attributed to subsets of Bonami [49], Gross [125], Beckner [26], each of whom has worked on estimates of this type. The proof is omitted.

4.47 Proposition. *For $g : \Omega \to \mathbb{R}$ and $\rho > 0$,*

$$\|T_\rho g\|_2 \leq \|g\|_{1+\rho^2}.$$

Let $0 < \rho < 1$. Set $g = f_e$ where $f = 1_A$, noting that g takes the values $0, \pm 1$ only. Then,

$$\sum_{F: e \in F} 4 \hat{f}(F)^2 \rho^{2|F|}$$

$$= \sum_{F \subseteq E} \hat{f_e}(F)^2 \rho^{2|F|} \qquad\qquad \text{by (4.43)}$$

$$= \| T_\rho f_e \|_2^2$$

$$\leq \| f_e \|_{1+\rho^2}^2 = \big[\phi(|f_e|^{1+\rho^2}) \big]^{2/(1+\rho^2)} \quad \text{by Proposition 4.47}$$

$$= \| f_e \|_2^{4/(1+\rho^2)} = I(e)^{2/(1+\rho^2)} \qquad \text{by (4.44).}$$

Therefore,

(4.48) $$\sum_{e \in E} I(e)^{2/(1+\rho^2)} \geq 4 \sum_{F \subseteq E} |F| \hat{f}(F)^2 \rho^{2|F|}.$$

Let $t = \phi(A) = \hat{f}(\varnothing)$. By (4.48),

(4.49) $$\sum_{e \in E} I(e)^{2/(1+\rho^2)} \geq 4 \rho^{2b} \sum_{0 < |F| \leq b} \hat{f}(F)^2$$

$$= 4 \rho^{2b} \bigg(\sum_{|F| \leq b} \hat{f}(F)^2 - t^2 \bigg),$$

where $b \in (0, \infty)$ will be chosen later. By (4.45),

$$\sum_{e \in E} I(e) \geq 4b \sum_{|F| > b} \hat{f}(F)^2,$$

which we add to (4.49) to obtain

(4.50) $$\rho^{-2b} \sum_{e \in E} I(e)^{2/(1+\rho^2)} + \frac{1}{b} \sum_{e \in E} I(e)$$

$$\geq 4 \sum_{F \subseteq E} \hat{f}(F)^2 - 4t^2$$

$$= 4t(1 - t) \qquad \text{by (4.42).}$$

We are now ready to prove (4.30). Let $m = \max_e I(e)$, noting that $m > 0$ since $\phi(A) \neq 0, 1$. The claim is trivial if $m = 1$, and we assume that $m < 1$. Then

$$\sum_{e \in E} I(e)^{4/3} \leq m^{1/3} \sum_{e \in E} I(e),$$

whence, by (4.50) and the choice $\rho^2 = \frac{1}{2}$,

$$(4.51) \qquad \left(2^b m^{1/3} + \frac{1}{b}\right) \sum_{e \in E} I(e) \geq 4t(1-t).$$

We choose b such that $2^b m^{1/3} = b^{-1}$, and it is an easy exercise that $b \geq A \log(1/m)$ for some absolute constant $A > 0$. With this choice of b, (4.30) follows from (4.51) with $c = 2A$. Inequality (4.35) follows, as explained after the statement of the theorem. $\qquad \square$

Proof of Theorem 4.33. We follow [92]. The idea of the proof is to 'discretize' the cube K and the increasing event A, and to apply Theorem 4.29.

Let $k \in \{1, 2, \dots\}$ to be chosen later, and partition the N-cube $K = [0, 1]^E$ into 2^{kN} disjoint smaller cubes each of side-length 2^{-k}. These small cubes are of the form

$$(4.52) \qquad B(\mathbf{l}) = \prod_{e \in E} [l_e, l_e + 2^{-k}),$$

where $\mathbf{l} = (l_e : e \in E)$ and each l_e is a 'binary decimal' of the form $l_e = 0.l_{e,1} l_{e,2} \cdots l_{e,k}$ with each $l_{e,j} \in \{0, 1\}$. There is a special case. When $l_e = 0.11 \cdots 1$, we put the *closed* interval $[l_e, l_e + 2^{-k}]$ into the product of (4.52). Lebesgue measure λ on K induces product measure ϕ with density $\frac{1}{2}$ on the space $\Omega = \{0, 1\}^{kN}$ of 0/1-vectors $(l_{e,j} : j = 1, 2, \dots, k, e \in E)$. We call each $B(\mathbf{l})$ a 'small cube'.

We claim that it suffices to consider events A that are the unions of small cubes. For a measurable subset $A \subseteq K$, let $\hat{A}$ be the subset of K that 'approximates' to A, given by $\hat{A} = \bigcup_{\mathbf{l} \in \mathcal{A}} B(\mathbf{l})$, where

$$\mathcal{A} = \{\mathbf{l} \in \Omega : B(\mathbf{l}) \cap A \neq \varnothing\}.$$

Note that $\mathcal{A}$ is an increasing subset of the discrete kN-cube Ω. We write $I_{\mathcal{A}}(e, j)$ for the influence of the index (e, j) on the subset $\mathcal{A} \subseteq \Omega$ under the measure ϕ. The next task is to show that, when replacing A by $\hat{A}$, the measure and influences of A are not greatly changed.

4.53 Lemma [51]. *In the above notation,*

$$(4.54) \qquad 0 \leq \lambda(\hat{A}) - \lambda(A) \leq \frac{N}{2^k},$$

$$(4.55) \qquad |I_{\hat{A}}(e) - I_A(e)| \leq \frac{2N}{2^k}, \qquad e \in E.$$

Proof. Clearly $A \subseteq \hat{A}$, whence $\lambda(A) \leq \lambda(\hat{A})$. Let $\mu : K \to K$ be the projection mapping that maps $(x_f : f \in E)$ to $(x_f - m : f \in E)$, where

$m = \min_{g \in E} x_g$. We have that

$$(4.56) \qquad \lambda(\hat{A}) - \lambda(A) \le |R| 2^{-kN},$$

where R is the set of small cubes that intersect both A and its complement $\bar{A}$. Since A is increasing, R cannot contain two distinct elements r, r' with $\mu(r) = \mu(r')$. Therefore, $|R|$ is no larger than the number of faces of small cubes lying in the 'hyperfaces' of K, that is,

$$(4.57) \qquad K \le N 2^{k(N-1)}.$$

Inequality (4.54) follows by (4.56).

Let $e \in E$, and let $\mu_e : K \to [0, 1]^{E \setminus \{e\}}$ be the projection that sends $(x_f : f \in E)$ to $(x_f : f \in E \setminus \{e\})$. The face $\mu_e(K)$ is the union of 'small faces' of small cubes. Each small face F corresponds to a 'tube' $T(F)$ of small cubes, based on that face with axis parallel to the eth direction (see Figure 4.1). Such a tube has 'first' face F and 'last' face $L = T(F) \cap \{\omega \in K : \omega(e) = 1\}$, and we write B_F (respectively, B_L) for the (unique) small cube with face F (respectively, L).

It is easily seen that F contributes 0 to $I_{\hat{A}}(e) - I_A(e)$ if 1_A is constant on both B_F and B_L (it is not important that 1_A should take the same value on the initial small cube as on the final). Therefore,

$$(4.58) \qquad |I_{\hat{A}}(e) - I_A(e)| \le |N_F \cup N_L| 2^{-k(N-1)},$$

where N_F (respectively, N_L) is the set of initial (respectively, final) small cubes on which 1_A is non-constant. By restricting 1_A to the 'fattened hyperface' $\bigcup \{B_F : F \subseteq \mu_e(K)\}$ and applying the argument leading to (4.57) within this region, we find as there that

$$|N_F| \le (N - 1) 2^{k(N-2)}.$$

The same inequality holds with N_L in place of N_F, and inequality (4.55) follows by (4.58). $\qquad \square$

Let A be an increasing subset of K, assume $0 < t = \lambda(A) < 1$, and let $m = \max_e I_A(e)$. We may assume that $0 < m < \frac{1}{2}$, since otherwise (4.34) is a triviality. With $\hat{A}$ given as above for some value of k to be chosen soon, we write $\hat{t} = \lambda(\hat{A})$ and $\hat{m} = \max_e I_{\hat{A}}(e)$. We shall prove below that

$$(4.59) \qquad \sum_{e \in E} I_{\hat{A}}(e) \ge c \hat{t} (1 - \hat{t}) \log[1/(2\hat{m})],$$

for some absolute constant $c > 0$. Let $k = k(N, A)$ be sufficiently large

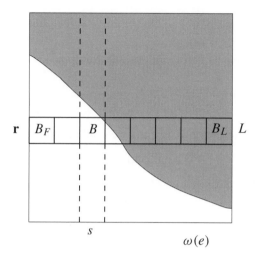

Figure 4.1. The small boxes $B = B(\mathbf{r}, s)$ form the tube $T(\mathbf{r})$. The region A is shaded.

that the following inequalities hold:

(4.60)
$$\frac{N}{2^k} < \frac{1}{2} \min\left\{ t(1-t), \frac{m \log[1/(2m)]}{2 + \log[1/(2m)]}, \frac{1}{2} - m \right\},$$

(4.61)
$$\frac{2N^2}{2^k} < \frac{1}{8} ct(1-t) \log[1/(2m)].$$

By Lemma 4.53,

(4.62)
$$|t - \hat{t}| \le \frac{N}{2^k}, \qquad |m - \hat{m}| \le \frac{2N}{2^k},$$

whence, by (4.60),
(4.63)
$$|t - \hat{t}| \le \tfrac{1}{2} t(1-t), \quad \hat{m} < \tfrac{1}{2}, \quad \frac{|m - \hat{m}|}{m \wedge \hat{m}} \le \tfrac{1}{2} \log[1/(2m)].$$

By Lemma 4.53 again,

$$\sum_{e \in A} I_A(e) \ge \sum_{e \in A} I_{\hat{A}}(e) - \frac{2N^2}{2^k}.$$

By (4.59), (4.61), (4.63), and a little elementary calculus,

$$\sum_{e \in A} I_A(e)$$

$$\geq c\Big[t(1-t) - |t - \hat{t}|\Big]\left[\log[1/(2m)] - \frac{|m - \hat{m}|}{m \wedge \hat{m}}\right] - \frac{2N^2}{2^k}.$$

$$\geq \tfrac{1}{8}ct(1-t)\log[1/(2m)]$$

as required.

It thus suffices to prove (4.59), and we shall henceforth assume that

(4.64) A is a union of small cubes.

4.65 Lemma [51, 92]. *For $e \in E$,*

$$\sum_{j=1}^{k} I_A(e, j) \leq 2I_A(e).$$

Proof. Let $e \in E$. For a fixed vector $\mathbf{r} = (r_1, r_2, \ldots, r_{N-1}) \in (\{0, 1\}^k)^{E \setminus \{e\}}$, consider the 'tube' $T(\mathbf{r})$ comprising the union of the small cubes $B(\mathbf{r}, s)$ of (4.52) over the 2^k possible values in $s \in \{0, 1\}^k$. We see after a little thought (see Figure 4.1) that

$$I_A(e, j) = \sum_{\mathbf{r}} (\tfrac{1}{2})^{kN-1} K(\mathbf{r}, j),$$

where $K(\mathbf{r}, j)$ is the number of unordered pairs $S = B(\mathbf{r}, s)$, $S' = B(\mathbf{r}, s')$ of small cubes of $T(\mathbf{r})$ such that: $S \subseteq A$, $S' \not\subseteq A$, and $|s - s'| = 2^{-j}$. Since A is an increasing subset of K, we can see that

$$K(\mathbf{r}, j) \leq 2^{k-j}, \qquad j = 1, 2, \ldots, k,$$

whence

$$\sum_{j} I_A(e, j) \leq \frac{2^k}{2^{kN-1}} J_N = \frac{2}{2^{k(N-1)}} J_N,$$

where J_N is the number of tubes $T(\mathbf{r})$ that intersect both A and its complement $\bar{A}$. By (4.64),

$$I_A(e) = \frac{1}{2^{k(N-1)}} J_N,$$

and the lemma is proved.   □

We return to the proof of (4.59). Assume that $m = \max_e I_A(e) < \tfrac{1}{2}$. By Lemma 4.65,

$$I_A(e, j) \leq 2m \qquad \text{for all } e, j.$$

By (4.30) applied to the event $\mathcal{A}$ of the kN-cube Ω,

$$\sum_{e,j} I_{\mathcal{A}}(e, j) \geq c_1 t(1 - t) \log[1/(2m)],$$

where c_1 is an absolute positive constant and $t = \lambda(A)$. By Lemma 4.65 again,

$$\sum_{e \in E} I_A(e) \geq \tfrac{1}{2} c_1 t(1 - t) \log[1/(2m)],$$

as required at (4.59).   □

Proof of Theorem 4.38. We prove this in two steps.

  I. In the notation of the theorem, there exists a Lebesgue-measurable subset B of $K = [0, 1]^E$ such that: $\mathbb{P}(A) = \lambda(B)$, and $I_A(e) \geq I_B(e)$ for all e, where the influences are calculated according to the appropriate probability measures.

 II. There exists an increasing subset C of K such that $\lambda(B) = \lambda(C)$, and $I_B(e) \geq I_C(e)$ for all e.

The claims of the theorem follow via Theorem 4.33 from these two facts.

A version of Claim I was stated in [51] without proof. We use the measure-space isomorphism theorem, Theorem B of [126, p. 173] (see also [1, p. 3] or [199, p. 16]). Let $x_1, x_2, \ldots$ be an ordering of the atoms of X, and let Q_i be the sub-interval $[q_i, q_{i+1})$ of $[0, 1]$, where $q_1 = 0$ and

$$q_i = \sum_{j=1}^{i-1} P(\{x_j\}) \quad \text{for } i \geq 2, \qquad q_\infty = \sum_{j \geq 1} P(\{x_j\}).$$

The non-atomic part of X has sample space $\Sigma' = \Sigma \setminus \{x_1, x_2, \ldots\}$, and total measure $1 - q_\infty$. By the isomorphism theorem, there exists a measure-preserving map μ from the σ-algebra $\mathcal{F}'$ of Σ' to the Borel σ-algebra of the interval $[q_\infty, 1]$ endowed with Lebesgue measure λ_1, satisfying

$$\mu(A_1 \setminus A_2) \stackrel{\lambda}{=} \mu A_1 \setminus \mu A_2,$$

(4.66)
$$\mu\left(\bigcup_{n=1}^{\infty} A_n\right) \stackrel{\lambda}{=} \bigcup_{n=1}^{\infty} \mu A_n,$$

for $A_n \in \mathcal{F}'$, where $A \stackrel{\lambda}{=} B$ means that $\lambda_1(A \triangle B) = 0$. We extend the domain of μ to $\mathcal{F}$ by setting $\mu(\{x_i\}) = Q_i$. In summary, there exists $\mu : \mathcal{F} \to \mathcal{B}[0, 1]$ such that $P(A) = \lambda_1(\mu A)$ for $A \in \mathcal{F}$, and (4.66) holds for $A_n \in \mathcal{F}$.

The product σ-algebra $\mathcal{F}^E$ of X^E is generated by the class $\mathcal{R}^E$ of 'rectangles' of the form $R = \prod_{e \in E} A_e$ for $A_e \in \mathcal{F}$. For such $R \in \mathcal{R}^E$, let

$$\mu^E R = \prod_{e \in E} \mu A_e.$$

We extend the domain of μ^E to the class $\mathcal{U}$ of finite unions of rectangles by

$$\mu^E \left(\bigcup_{i=1}^m R_i \right) = \bigcup_{i=1}^m \mu^E R_i.$$

It can be checked that

(4.67) $$\mathbb{P}(R) = \lambda(\mu^E R),$$

for any such union R.

Let $A \in \mathcal{F}^E$, and assume without loss of generality that $0 < \lambda(A) < 1$. We can find an increasing sequence $(U_n : n \geq 1)$ of elements of $\mathcal{U}$, each being a finite union of rectangles with strictly positive measure, such that $\mathbb{P}(A \triangle U_n) \to 0$ as $n \to \infty$, and in particular

(4.68) $$\mathbb{P}(U_n \setminus A) = 0, \qquad n \geq 1.$$

Let $V_n = \mu^E U_n$ and $B = \lim_{n \to \infty} V_n$. Since V_n is non-decreasing in n, by (4.67),

$$\lambda(B) = \lim_{n \to \infty} \lambda(\mu^E U_n) = \lim_{n \to \infty} \mathbb{P}(U_n) = \mathbb{P}(A).$$

We turn now to the influences. Let $e \in E$. For $\psi \in \Sigma^{E \setminus \{e\}}$, let $F_\psi = \{\psi\} \times \Sigma$ be the 'fibre' at ψ, and

$$J_A^a = P^{E \setminus \{e\}} \left(\{ \psi \in \Sigma^{E \setminus \{e\}} : P(A \cap F_\psi) = a \} \right), \qquad a = 0, 1.$$

We define J_B^a similarly, with P replaced by λ and Σ replaced by $[0, 1]$. Thus,

(4.69) $$I_A(e) = 1 - J_A^0 - J_A^1,$$

and we claim that

(4.70) $$J_A^0 \leq J_B^0.$$

By replacing A by its complement $\overline{A}$, we obtain that $J_A^1 \leq J_B^1$, and it follows by (4.69)–(4.70) that $I_A(e) \geq I_B(e)$, as required. We write U_n as the finite union $U_n = \bigcup_i F_i \times G_i$, where each F_i (respectively, G_i) is a rectangle of $\Sigma^{E \setminus \{e\}}$ (respectively, Σ). By Fubini's theorem and (4.68),

$$J_A^0 \leq J_{U_n}^0 = 1 - P^{E \setminus \{e\}} \left(\bigcup_i F_i \right)$$

$$= 1 - \lambda^{E \setminus \{e\}} \left(\bigcup_i \mu^{E \setminus \{e\}} F_i \right) = J_{V_n}^0,$$

by (4.67) with E replaced by $E \setminus \{e\}$.

Finally, we show that $J^0_{V_n} \to J^0_B$ as $n \to \infty$, and (4.70) will follow. For $\psi \in \Sigma^E$, we write $\mathrm{proj}(\psi)$ for the projection of ψ onto the sub-space $\Sigma^{E \setminus \{e\}}$. Since the V_n are unions of rectangles of $[0, 1]^E$ with strictly positive measure,

$$J^0_{V_n} = \lambda^{E \setminus \{e\}} \left(\mathrm{proj}\, \overline{V_n} \right).$$

Now, $V_n \uparrow B$, so that $\omega \in B$ if and only if $\omega \in V_n$ for some n. It follows that $\mathrm{proj}\, \overline{V_n} \downarrow \mathrm{proj}\, \overline{B}$, whence $J^0_{V_n} \to \lambda^{E \setminus \{e\}}(\mathrm{proj}\, \overline{B})$. Also, $\lambda^{E \setminus \{e\}}(\mathrm{proj}\, \overline{B}) = J^0_B$, and (4.70) follows. Claim I is proved.

Claim II is proved by an elaboration of the method laid out in [30, 51]. Let $B \subseteq K$ be a non-increasing event. For $e \in E$ and $\psi = (\omega(g) : g \neq e) \in [0, 1]^{E \setminus \{e\}}$, we define the fibre F_ψ as usual by $F_\psi = \{\psi\} \times [0, 1]$. We replace $B \cap F_\psi$ by the set

(4.71)
$$B_\psi = \begin{cases} \{\psi\} \times (1 - y, 1] & \text{if } y > 0, \\ \varnothing & \text{if } y = 0, \end{cases}$$

where

(4.72)
$$y = y(\psi) = \lambda_1(B \cap F_\psi).$$

Thus B_ψ is obtained from B by 'pushing $B \cap F_\psi$ up the fibre' in a measure-preserving manner (see Figure 4.2). Clearly, $M_e B = \bigcup_\psi B_\psi$ is increasing[4] in the direction e and, by Fubini's theorem,

(4.73)
$$\lambda(M_e B) = \lambda(B).$$

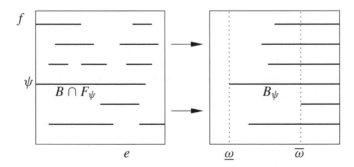

Figure 4.2. In the e/f-plane, we push every $B \cap F_\psi$ as far rightwards along the fibre F_ψ as possible.

[4]*Exercise*: Show that $M_e B$ is Lebesgue-measurable.

We order E in an arbitrary manner, and let

$$C = \left(\prod_{e \in E} M_e \right) B,$$

where the product is constructed in the given order. By (4.73), $\lambda(C) = \lambda(B)$. We show that C is increasing by proving that: if B is increasing in direction $f \in E$, where $f \neq e$, then so is $M_e B$. It is enough to work with the reduced sample space $K' = [0, 1]^{\{e, f\}}$, as illustrated in Figure 4.2. Suppose that $\omega, \omega' \in K'$ are such that $\omega(e) = \omega'(e)$ and $\omega(f) < \omega'(f)$. Then

(4.74) $$1_{M_e B}(\omega) = \begin{cases} 1 & \text{if } \omega(e) > 1 - y, \\ 0 & \text{if } \omega(e) \leq 1 - y, \end{cases}$$

where $y = y(\omega(f))$ is given according to (4.72), with a similar expression with ω and y replaced by ω' and y'. Since B is assumed increasing in $\omega(f)$, we have that $y \leq y'$. By (4.74), if $\omega \in M_e B$, then $\omega' \in M_e B$, which is to say that $M_e B$ is increasing in direction f.

Finally, we show that

(4.75) $$I_{M_e B}(f) \leq I_B(f), \qquad f \in E,$$

whence $I_C(f) \leq I_B(f)$ and the theorem is proved. First, by construction, $I_{M_e B}(e) = I_B(e)$. Let $f \neq e$. By conditioning on $\omega(g)$ for $g \neq e, f$,

$$I_{M_e B}(f) = \lambda^{E \setminus \{e, f\}} \Big(\lambda_1 \big(\{\omega(e) : 0 < \lambda_1(M_e B \cap F_v) < 1\} \big) \Big),$$

where $v = (\omega(g) : g \neq f)$ and $F_v = \{v\} \times [0, 1]$. We shall show that

(4.76) $$\lambda_1 \big(\{\omega(e) : 0 < \lambda_1(M_e B \cap F_v) < 1\} \big)$$
$$\leq \lambda_1 \big(\{\omega(e) : 0 < \lambda_1(B \cap F_v) < 1\} \big),$$

and the claim will follow. Inequality (4.76) depends only on $\omega(e)$, $\omega(f)$, and thus we shall make no further reference to the remaining coordinates $\omega(g)$, $g \neq e, f$. Henceforth, we write ω for $\omega(e)$ and ψ for $\omega(f)$.

With the aid of Figure 4.2, we see that the left side of (4.76) equals $\overline{\omega} - \underline{\omega}$, where

(4.77) $$\overline{\omega} = \sup\{\omega : \lambda_1(M_e B \cap F_\omega) < 1\},$$
$$\underline{\omega} = \inf\{\omega : \lambda_1(M_e B \cap F_\omega) > 0\}.$$

We may assume that $\underline{\omega} < 1$ and $\overline{\omega} > 0$, since otherwise $\overline{\omega} = \underline{\omega}$ and (4.76) is trivial. Let ϵ be positive and small, and let

(4.78) $$A_\epsilon = \{\psi : \lambda_1(B \cap F_\psi) > 1 - \underline{\omega} - \epsilon\}.$$

Since $\lambda_1(B \cap F_\psi) = \lambda_1(M_e B \cap F_\psi)$, $\lambda_1(A_\epsilon) > 0$ by (4.77). Let $A'_\epsilon = [0, 1] \times A_\epsilon$. We now estimate the two-dimensional Lebesgue measure $\lambda_2(B \cap A'_\epsilon)$ in two ways:

$$\lambda_2(B \cap A'_\epsilon) > \lambda_1(A_\epsilon)(1 - \underline{\omega} - \epsilon) \qquad \text{by (4.78),}$$

$$\lambda_2(B \cap A'_\epsilon) \leq \lambda_1(A_\epsilon)\lambda_1(\{\omega : \lambda_1(B \cap F_\omega) > 0\}),$$

whence $D_0 = \{\omega : \lambda_1(B \cap F_\omega) > 0\}$ satisfies

$$\lambda_1(D_0) \geq \lim_{\epsilon \downarrow 0}[1 - \underline{\omega} - \epsilon] = 1 - \underline{\omega}.$$

By a similar argument, $D_1 = \{\omega : \lambda_1(B \cap F_\omega) = 1\}$ satisfies

$$\lambda_1(D_1) \leq 1 - \overline{\omega}.$$

For $\omega \in D_0 \setminus D_1$, $0 < \lambda_1(B \cap F_\omega) < 1$, so that

$$I_B(e) \geq \lambda_1(D_0 \setminus D_1) \geq \overline{\omega} - \underline{\omega},$$

and (4.75) follows. □

4.7 Russo's formula and sharp thresholds

Let ϕ_p denote product measure with density p on the finite product space $\Omega = \{0, 1\}^E$. The influence $I_A(e)$, of $e \in E$ on an event A, is given in (4.28).

4.79 Theorem (Russo's formula). *For any event $A \subseteq \Omega$,*

$$\frac{d}{dp}\phi_p(A) = \sum_{e \in E}[\phi_p(A^e) - \phi(A_e)] = \sum_{e \in E} I_A(e).$$

This formula, or its equivalent, has been discovered by a number of authors. See, for example, [24, 184, 212]. The element $e \in E$ is called *pivotal* for the event A if the occurrence or not of A depends on the state of e, that is, if $1_A(\omega_e) \neq 1_A(\omega^e)$. If A is increasing, Russo's formula states that $\phi'_p(A)$ equals the mean number of pivotal elements of E.

Proof. This is standard, see for example [106]. Since

$$\phi_p(A) = \sum_\omega 1_A(\omega)\phi_p(\omega),$$

it is elementary that

$$(4.80) \qquad \frac{d}{dp}\phi_p(A) = \sum_{\omega \in \Omega}\left(\frac{|\eta(\omega)|}{p} - \frac{N - |\eta(\omega)|}{1 - p}\right)1_A(\omega)\phi_p(\omega),$$

where $\eta(\omega) = \{e \in E : \omega(e) = 1\}$ and $N = |E|$. Let 1_e be the indicator function that e is open. Since $\phi_p(1_e) = p$ for all $e \in E$, and $|\eta| = \sum_e 1_e$,

$$p(1 - p)\frac{d}{dp}\phi_p(A) = \phi_p([|\eta| - pN]1_A)$$

$$= \sum_{e \in E}[\phi_p(1_e 1_A) - \phi_p(1_e)\phi_p(1_A)].$$

The summand equals

$$p\phi_p(A^e) - p[p\phi_p(A^e) + (1 - p)\phi_p(A_e)],$$

and the formula is proved.   □

Let A be an increasing subset of $\Omega = \{0, 1\}^E$ that is non-trivial in that $A \neq \varnothing, \Omega$. The function $f(p) = \phi_p(A)$ is non-decreasing with $f(0) = 0$ and $f(1) = 1$. The next theorem is an immediate consequence of Theorems 4.38 and 4.79.

4.81 Theorem [231]. *There exists a constant $c > 0$ such that the following holds. Let A be an increasing subset of Ω with $A \neq \varnothing, \Omega$. For $p \in (0, 1)$,*

$$\frac{d}{dp}\phi_p(A) \geq c\phi_p(A)(1 - \phi_p(A)) \log[1/(2 \max_e I_A(e))],$$

where $I_A(e)$ is the influence of e on A with respect to the measure ϕ_p.

Theorem 4.81 takes an especially simple form when A has a certain property of symmetry. In such a case, the following sharp-threshold theorem implies that $f(p) = \phi_p(A)$ increases from (near) 0 to (near) 1 over an interval of p-values with length of order not exceeding $1/\log N$.

Let Π be the group of permutations of E. Any $\pi \in \Pi$ acts on Ω by $\pi\omega = (\omega(\pi_e) : e \in E)$. We say that a subgroup $\mathcal{A}$ of Π acts *transitively* on E if, for all pairs $j, k \in E$, there exists $\alpha \in \mathcal{A}$ with $\alpha_j = k$.

Let $\mathcal{A}$ be a subgroup of Π. A probability measure ϕ on $(\Omega, \mathcal{F})$ is called $\mathcal{A}$-*invariant* if $\phi(\omega) = \phi(\alpha\omega)$ for all $\alpha \in \mathcal{A}$. An event $A \in \mathcal{F}$ is called $\mathcal{A}$-*invariant* if $A = \alpha A$ for all $\alpha \in \mathcal{A}$. It is easily seen that, for any subgroup $\mathcal{A}$, ϕ_p is $\mathcal{A}$-invariant.

4.82 Theorem (Sharp threshold) [93]. *There exists a constant c satisfying $c \in (0, \infty)$ such that the following holds. Let $N = |E| \geq 1$. Let $A \in \mathcal{F}$ be an increasing event, and suppose there exists a subgroup $\mathcal{A}$ of Π acting transitively on E such that A is $\mathcal{A}$-invariant. Then*

$$(4.83) \qquad \frac{d}{dp}\phi_p(A) \geq c\phi_p(A)(1 - \phi_p(A)) \log N, \qquad p \in (0, 1).$$

Proof. We show first that the influences $I_A(e)$ are constant for $e \in E$. Let $e, f \in E$, and find $\alpha \in \mathcal{A}$ such that $\alpha_e = f$. Under the given conditions,

$$\phi_p(A, 1_f = 1) = \sum_{\omega \in A} \phi_p(\omega) 1_f(\omega) = \sum_{\omega \in A} \phi_p(\alpha\omega) 1_e(\alpha\omega)$$

$$= \sum_{\omega' \in A} \phi_p(\omega') 1_e(\omega') = \phi_p(A, 1_e = 1),$$

where 1_g is the indicator function that $\omega(g) = 1$. On setting $A = \Omega$, we deduce that $\phi_p(1_f = 1) = \phi_p(1_e = 1)$. On dividing, we obtain that $\phi_p(A \mid 1_f = 1) = \phi_p(A \mid 1_e = 1)$. A similar equality holds with 1 replaced by 0, and therefore $I_A(e) = I_A(f)$.

It follows that

$$\sum_{f \in E} I_A(f) = N I_A(e).$$

By Theorem 4.38 applied to the product space $(\Omega, \mathcal{F}, \phi_p)$, the right side is at least $c\phi_p(A)(1 - \phi_p(A)) \log N$, and (4.83) is a consequence of Theorem 4.79. □

Let $\epsilon \in (0, \frac{1}{2})$ and let A be increasing and non-trivial. Under the conditions of Theorem 4.82, $\phi_p(A)$ increases from ϵ to $1 - \epsilon$ over an interval of values of p having length of order not exceeding $1/\log N$. This amounts to a quantification of the so-called S-shape results described and cited in [106, Sect. 2.5]. An early step in the direction of sharp thresholds was taken by Russo [213] (see also [231]), but without the quantification of $\log N$.

Essentially the same conclusions hold for a family $\{\mu_p : p \in (0, 1)\}$ of probability measures given as follows in terms of a positive measure μ satisfying the FKG lattice condition. For $p \in (0, 1)$, let μ_p be given by

$$(4.84) \qquad \mu_p(\omega) = \frac{1}{Z_p} \left(\prod_{e \in E} p^{\omega(e)} (1 - p)^{1 - \omega(e)} \right) \mu(\omega), \qquad \omega \in \Omega,$$

where Z_p is chosen in such a way that μ_p is a probability measure. It is easy to check that each μ_p satisfies the FKG lattice condition. It turns out that, for an increasing event $A \neq \varnothing, \Omega$,

$$(4.85) \qquad \frac{d}{dp} \mu_p(A) \geq \frac{c\xi_p}{p(1 - p)} \mu_p(A)(1 - \mu_p(A)) \log[1/(2 \max_e J_A(e))],$$

where

$$\xi_p = \min_{e \in E} \left[\mu_p(\omega(e) = 1) \mu_p(\omega(e) = 0) \right].$$

The proof uses inequality (4.36), see [100, 101]. This extension of Theorem 4.81 does not appear to have been noted before. It may be used in the studies of the random-cluster model, and of the Ising model with external field (see [101]).

A slight variant of Theorem 4.82 is valid for measures ϕ_p given by (4.84), with the positive probability measure μ satisfying: μ satisfies the FKG lattice condition, and μ is $\mathcal{A}$-invariant. See (4.85) and [100, 109].

From amongst the issues arising from the sharp-threshold Theorem 4.82, we identify two. First, to what degree is information about the group $\mathcal{A}$ relevant to the sharpness of the threshold? Secondly, what can be said when $p = p_N$ tends to 0 as $N \to \infty$? The reader is referred to [147] for some answers to these questions.

4.8 Exercises

4.1 Let $X_n, Y_n \in L^2(\Omega, \mathcal{F}, \mathbb{P})$ be such that $X_n \to X$, $Y_n \to Y$ in L^2. Show that $X_n Y_n \to XY$ in L^1. [Reminder: L^p is the set of random variables Z with $\mathbb{E}(|Z|^p) < \infty$, and $Z_n \to Z$ in L^p if $\mathbb{E}(|Z_n - Z|^p) \to 0$. You may use any standard fact such as the Cauchy–Schwarz inequality.]

4.2 [135] Let $\mathbb{P}_p$ be the product measure on the space $\{0, 1\}^n$ with density p. Show by induction on n that $\mathbb{P}_p$ satisfies the Harris–FKG inequality, which is to say that $\mathbb{P}_p(A \cap B) \geq \mathbb{P}_p(A)\mathbb{P}_p(B)$ for any pair A, B of increasing events.

4.3 (continuation) Consider bond percolation on the square lattice $\mathbb{Z}^2$. Let X and Y be increasing functions on the sample space, such that $\mathbb{E}_p(X^2)$, $\mathbb{E}_p(Y^2) < \infty$. Show that X and Y are positively associated.

4.4 *Coupling.*

(a) Take $\Omega = [0, 1]$, with the Borel σ-field and Lebesgue measure $\mathbb{P}$. For any distribution function F, define a random variable Z_F on Ω by

$$Z_F(\omega) = \inf\{z : \omega \leq F(z)\}, \qquad \omega \in \Omega.$$

Prove that

$$\mathbb{P}(Z_F \leq z) = \mathbb{P}\big([0, F(z)]\big) = F(z),$$

whence Z_F has distribution function F.

(b) For real-valued random variables X, Y, we write $X \leq_{\mathrm{st}} Y$ if $\mathbb{P}(X \leq u) \geq \mathbb{P}(Y \leq u)$ for all u. Show that $X \leq_{\mathrm{st}} Y$ if and only if there exist random variables X', Y' on Ω, with the same respective distributions as X and Y, such that $\mathbb{P}(X' \leq Y') = 1$.

4.5 [109] Let μ be a positive probability measure on the finite product space $\Omega = \{0, 1\}^E$.

(a) Show that μ satisfies the FKG lattice condition

$$\mu(\omega_1 \vee \omega_2)\mu(\omega_1 \wedge \omega_2) \geq \mu(\omega_1)\mu(\omega_2), \qquad \omega_1, \omega_2 \in \Omega,$$

if and only if this inequality holds for all pairs ω_1, ω_2 that differ on exactly two elements of E.

(b) Show that the FKG lattice condition is equivalent to the statement that μ is *monotone*, in that, for $e \in E$,

$$f(e, \xi) := \mu\big(\omega(e) = 1 \,\big|\, \omega(f) = \xi(f) \text{ for } f \neq e\big)$$

is non-decreasing in $\xi \in \{0, 1\}^{E \setminus \{e\}}$.

4.6 [109] Let μ_1, μ_2 be positive probability measures on the finite product $\Omega = \{0, 1\}^E$. Assume that they satisfy

$$\mu_2(\omega_1 \vee \omega_2)\mu_1(\omega_1 \wedge \omega_2) \geq \mu_1(\omega_1)\mu_2(\omega_2),$$

for all pairs $\omega_1, \omega_2 \in \Omega$ that differ on exactly one element of E, and in addition that either μ_1 or μ_2 satisfies the FKG lattice condition. Show that $\mu_2 \geq_{\mathrm{st}} \mu_1$.

4.7 Let $X_1, X_2, \ldots$ be independent Bernoulli random variables with parameter p, and $S_n = X_1 + X_2 + \cdots + X_n$. Show by Hoeffding's inequality or otherwise that

$$\mathbb{P}\big(|S_n - np| \geq x\sqrt{n}\big) \leq 2\exp(-\tfrac{1}{2}x^2/m^2), \qquad x > 0,$$

where $m = \max\{p, 1 - p\}$.

4.8 Let $G_{n,p}$ be the random graph with vertex set $V = \{1, 2, \ldots, n\}$ obtained by joining each pair of distinct vertices by an edge with probability p (different pairs are joined independently). Show that the chromatic number $\chi_{n,p}$ satisfies

$$\mathbb{P}\big(|\chi_{n,p} - \mathbb{E}\chi_{n,p}| \geq x\big) \leq 2\exp(-\tfrac{1}{2}x^2/n), \qquad x > 0.$$

4.9 *Russo's formula.* Let X be a random variable on the finite sample space $\Omega = \{0, 1\}^E$. Show that

$$\frac{d}{dp}\mathbb{E}_p(X) = \sum_{e \in E} \mathbb{E}_p(\delta_e X),$$

where $\delta_e X(\omega) = X(\omega^e) - X(\omega_e)$, and ω^e (respectively, ω_e) is the configuration obtained from ω by replacing $\omega(e)$ by 1 (respectively, 0).

Let A be an increasing event, with indicator function 1_A. An edge e is called *pivotal* for the event A in the configuration ω if $\delta_e I_A(\omega) = 1$. Show that the derivative of $\mathbb{P}_p(A)$ equals the mean number of pivotal edges for A. Find a related formula for the second derivative of $\mathbb{P}_p(A)$.

What can you show for the third derivative, and so on?

4.10 [100] Show that every increasing subset of the cube $[0, 1]^N$ is Lebesgue-measurable.

4.11 Heads turn up with probability p on each of N coin flips. Let A be an increasing event, and suppose there exists a subgroup $\mathcal{A}$ of permutations of $\{1, 2, \ldots, N\}$ acting transitively, such that A is $\mathcal{A}$-invariant. Let p_c be the value of p such that $\mathbb{P}_p(A) = \frac{1}{2}$. Show that there exists an absolute constant $c > 0$ such that

$$\mathbb{P}_p(A) \geq 1 - N^{-c(p-p_c)}, \qquad p \geq p_c,$$

with a similar inequality for $p \leq p_c$.

4.12 Let μ be a positive measure on $\Omega = \{0, 1\}^E$ satisfying the FKG lattice condition. For $p \in (0, 1)$, let μ_p be the probability measure given by

$$\mu_p(\omega) = \frac{1}{Z_p} \left(\prod_{e \in E} p^{\omega(e)} (1 - p)^{1 - \omega(e)} \right) \mu(\omega), \qquad \omega \in \Omega.$$

Let A be an increasing event. Show that there exists an absolute constant $c > 0$ such that

$$\mu_{p_1}(A)[1 - \mu_{p_2}(A)] \leq \lambda^{B(p_2 - p_1)}, \qquad 0 < p_1 < p_2 < 1,$$

where

$$B = \inf_{p \in (p_1, p_2)} \left\{ \frac{c\xi_p}{p(1 - p)} \right\}, \qquad \xi_p = \min_{e \in E} \left[\mu_p(\omega(e) = 1) \mu_p(\omega(e) = 0) \right],$$

and λ satisfies

$$2 \max_{e \in E} J_A(e) \leq \lambda, \qquad e \in E, \ p \in (p_1, p_2),$$

with $J_A(e)$ the conditional influence of e on A.

5

Further percolation

The subcritical and supercritical phases of percolation are characterized respectively by the absence and presence of an infinite open cluster. Connection probabilities decay exponentially when $p < p_c$, and there is a unique infinite cluster when $p > p_c$. There is a power-law singularity at the point of phase transition. It is shown that $p_c = \frac{1}{2}$ for bond percolation on the square lattice. The Russo–Seymour–Welsh (RSW) method is described for site percolation on the triangular lattice, and this leads to a statement and proof of Cardy's formula.

5.1 Subcritical phase

In language borrowed from the theory of branching processes, a percolation process is termed *subcritical* if $p < p_c$, and *supercritical* if $p > p_c$.

In the subcritical phase, all open clusters are (almost surely) finite. The chance of a long-range connection is small, and it approaches zero as the distance between the endpoints diverges. The process is considered to be 'disordered', and the probabilities of long-range connectivities tend to zero *exponentially* in the distance. Exponential decay may be proved by elementary means for sufficiently small p, as in the proof of Theorem 3.2, for example. It is quite another matter to prove exponential decay for all $p < p_c$, and this was achieved for percolation by Aizenman and Barsky [6] and Menshikov [189, 190] around 1986.

The methods of Sections 5.1–5.4 are fairly robust with respect to choice of process and lattice. For concreteness, we consider bond percolation on $\mathbb{L}^d$ with $d \geq 2$. The first principal result is the following theorem, in which $\Lambda(n) = [-n, n]^d$ and $\partial \Lambda(n) = \Lambda(n) \setminus \Lambda(n-1)$.

5.1 Theorem [6, 189, 190]. *There exists $\psi(p)$, satisfying $\psi(p) > 0$ when $0 < p < p_c$, such that*

$$(5.2) \qquad \mathbb{P}_p(0 \leftrightarrow \partial \Lambda(n)) \leq e^{-n\psi(p)}, \qquad n \geq 1.$$

The reader is referred to [106] for a full account of this important theorem. The two proofs of Aizenman–Barsky and Menshikov have some interesting similarities, while differing in fundamental ways. An outline of Menshikov's proof is presented later in this section. The Aizenman–Barsky proof proceeds via an intermediate result, namely the following of Hammersley [128]. Recall the open cluster C at the origin.

5.3 Theorem [128]. *Suppose that* $\chi(p) = \mathbb{E}_p|C| < \infty$. *There exists* $\sigma(p) > 0$ *such that*

$$(5.4) \qquad \mathbb{P}_p(0 \leftrightarrow \partial \Lambda(n)) \leq e^{-n\sigma(p)}, \qquad n \geq 1.$$

Seen in the light of Theorem 5.1, we may take the condition $\chi(p) < \infty$ as a characterization of the subcritical phase. It is not difficult to see, using subadditivity, that the limit of $n^{-1} \log \mathbb{P}_p(0 \leftrightarrow \partial \Lambda(n))$ exists as $n \to \infty$. See [106, Thm 6.10].

Proof. Let $x \in \partial \Lambda(n)$, and let $\tau_p(0, x) = \mathbb{P}_p(0 \leftrightarrow x)$ be the probability that there exists an open path of $\mathbb{L}^d$ joining the origin to x. Let R_n be the number of vertices $x \in \partial \Lambda(n)$ with this property, so that the mean value of R_n is

$$(5.5) \qquad \mathbb{E}_p(R_n) = \sum_{x \in \partial \Lambda(n)} \tau_p(0, x).$$

Note that

$$(5.6) \qquad \sum_{n=0}^{\infty} \mathbb{E}_p(R_n) = \sum_{n=0}^{\infty} \sum_{x \in \partial \Lambda(n)} \tau_p(0, x)$$

$$= \sum_{x \in \mathbb{Z}^d} \tau_p(0, x)$$

$$= \mathbb{E}_p|\{x \in \mathbb{Z}^d : 0 \leftrightarrow x\}| = \chi(p).$$

If there exists an open path from the origin to some vertex of $\partial \Lambda(m + k)$, then there exists a vertex x in $\partial \Lambda(m)$ that is connected by *disjoint* open paths both to the origin and to a vertex on the surface of the translate $\partial \Lambda(k, x) = x + \Lambda(k)$ (see Figure 5.1). By the BK inequality,

$$(5.7)$$

$$\mathbb{P}_p(0 \leftrightarrow \partial \Lambda(m + k)) \leq \sum_{x \in \partial \Lambda(m)} \mathbb{P}_p(0 \leftrightarrow x) \mathbb{P}_p(x \leftrightarrow x + \partial \Lambda(k))$$

$$= \sum_{x \in \partial \Lambda(m)} \tau_p(0, x) \mathbb{P}_p(0 \leftrightarrow \partial \Lambda(k)).$$

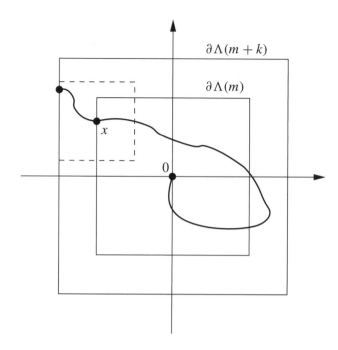

Figure 5.1. The vertex x is joined by disjoint open paths to the origin and to the surface of the translate $\Lambda(k, x) = x + \Lambda(k)$, indicated by the dashed lines.

by translation-invariance. Therefore,

$$(5.8) \quad \mathbb{P}_p(0 \leftrightarrow \partial\Lambda(m + k)) \le \mathbb{E}_p(R_m)\mathbb{P}_p(0 \leftrightarrow \partial\Lambda(k)), \qquad m, k \ge 1.$$

Whereas the BK inequality makes this calculation simple, Hammersley [128] employed a more elaborate argument by conditioning.

Let p be such that $\chi(p) < \infty$, so that $\sum_{m=0}^{\infty} \mathbb{E}_p(R_m) < \infty$ from (5.6). Then $\mathbb{E}_p(R_m) \to 0$ as $m \to \infty$, and we may choose m such that $\eta = \mathbb{E}_p(R_m)$ satisfies $\eta < 1$. Let n be a positive integer and write $n = mr + s$, where r and s are non-negative integers and $0 \le s < m$. Then

$$
\begin{aligned}
\mathbb{P}_p(0 \leftrightarrow \partial\Lambda(n)) &\le \mathbb{P}_p(0 \leftrightarrow \partial\Lambda(mr)) && \text{since } n \ge mr \\
&\le \eta^r && \text{by iteration of (5.8)} \\
&\le \eta^{-1+n/m} && \text{since } n < m(r + 1),
\end{aligned}
$$

which provides an exponentially decaying bound of the form of (5.4), valid for $n \ge m$. It is left as an exercise to extend the inequality to $n < m$. $\quad\square$

Outline proof of Theorem 5.1. The full proof can be found in [105, 106, 190, 237]. Let $S(n)$ be the 'diamond' $S(n) = \{x \in \mathbb{Z}^d : \delta(0, x) \le n\}$ containing all points within graph-theoretic distance n of the origin, and write $A_n = \{0 \leftrightarrow \partial S(n)\}$. We are concerned with the probabilities $g_p(n) = \mathbb{P}_p(A_n)$.

By Russo's formula, Theorem 4.79,

(5.9) $$g'_p(n) = \mathbb{E}_p(N_n),$$

where N_n is the number of pivotal edges for A_n, that is, the number of edges e for which $1_A(\omega^e) \ne 1_A(\omega_e)$. By a simple calculation,

(5.10) $$g'_p(n) = \frac{1}{p}\mathbb{E}_p(N_n 1_{A_n}) = \frac{1}{p}\mathbb{E}_p(N_n \mid A_n)g_p(n),$$

which may be integrated to obtain

(5.11) $$g_\alpha(n) = g_\beta(n) \exp\left(-\int_\alpha^\beta \frac{1}{p}\mathbb{E}_p(N_n \mid A_n)\,dp\right)$$
$$\le g_\beta(n) \exp\left(-\int_\alpha^\beta \mathbb{E}_p(N_n \mid A_n)\,dp\right),$$

where $0 < \alpha < \beta < 1$. The vast majority of the work in the proof is devoted to showing that $\mathbb{E}_p(N_n \mid A_n)$ grows at least linearly in n when $p < p_c$, and the conclusion of the theorem then follows immediately.

The rough argument is as follows. Let $p < p_c$, so that $\mathbb{P}_p(A_n) \to 0$ as $n \to \infty$. In calculating $\mathbb{E}_p(N_n \mid A_n)$, we are conditioning on an event of diminishing probability, and thus it is feasible that there are many pivotal edges of A_n. This will be proved by bounding (above) the mean distance between consecutive pivotal edges, and then applying a version of Wald's equation. The BK inequality, Theorem 4.17, plays an important role.

Suppose that A_n occurs, and denote by $e_1, e_2, \dots, e_N$ the pivotal edges for A_n in the order in which they are encountered when building the cluster from the origin. It is easily seen that all open paths from the origin to $\partial S(n)$ traverse every e_j. Furthermore, as illustrated in Figure 5.2, there must exist at least two edge-disjoint paths from the second endpoint of each e_j (in the above ordering) to the first of e_{j+1}.

Let $M = \max\{k : A_k \text{ occurs}\}$, so that

$$\mathbb{P}_p(M \ge k) = g_p(k) \to 0 \qquad \text{as } k \to \infty.$$

The key inequality states that

(5.12) $$\mathbb{P}_p(N_n \ge k \mid A_n) \ge \mathbb{P}(M_1 + M_2 + \cdots + M_k \le n - k),$$

where the M_i are independent copies of M. This is proved using the BK inequality, using the above observation concerning disjoint paths between

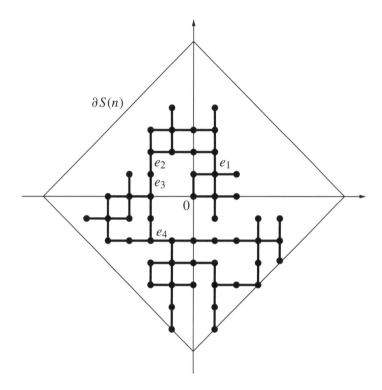

Figure 5.2. Assume that $0 \leftrightarrow \partial S(n)$. For any consecutive pair e_j, e_{j+1} of pivotal edges, taken in the order of traversal from 0 to $\partial S(n)$, there must exist at least two edge-disjoint open paths joining the second vertex of e_j and the first of e_{j+1}.

consecutive pivotal edges. The proof is omitted here. By (5.12),

$$(5.13) \qquad \mathbb{P}_p(N_n \geq k \mid A_n) \geq \mathbb{P}(M_1' + M_2' + \cdots + M_k' \leq n),$$

where $M_i' = 1 + \min\{M_i, n\}$. Summing (5.13) over k, we obtain

$$(5.14) \qquad \mathbb{E}_p(N_n \mid A_n) \geq \sum_{k=1}^{\infty} \mathbb{P}(M_1' + M_2' + \cdots + M_k' \leq n)$$

$$= \sum_{k=1}^{\infty} \mathbb{P}_p(K \geq k+1) = \mathbb{E}(K) - 1,$$

where $K = \min\{k : S_k > n\}$ and $S_k = M_1' + M_2' + \cdots + M_k'$. By Wald's equation,

$$n < \mathbb{E}(S_K) = \mathbb{E}(K)\mathbb{E}(M_1'),$$

whence

$$\mathbb{E}(K) > \frac{n}{\mathbb{E}(M_1')} = \frac{n}{1 + \mathbb{E}(\min\{M_1, n\})} = \frac{n}{\sum_{i=0}^{n} g_p(i)}.$$

In summary, this shows that

(5.15) $$\mathbb{E}_p(N_n \mid A_n) \geq \frac{n}{\sum_{i=0}^{n} g_p(i)} - 1, \qquad 0 < p < 1.$$

Inequality (5.15) may be fed into (5.10) to obtain a differential inequality for the $g_p(k)$. By a careful analysis of the latter inequality, we obtain that $\mathbb{E}_p(N_n \mid A_n)$ grows at least linearly with n whenever p satisfies $0 < p < p_c$. This step is neither short nor easy, but it is conceptually straightforward, and it completes the proof.   □

5.2 Supercritical phase

The critical value p_c is the value of p at which the percolation probability $\theta(p)$ becomes strictly positive. It is widely believed that $\theta(p_c) = 0$, and this is perhaps the major conjecture of the subject.

5.16 Conjecture. *For percolation on $\mathbb{L}^d$ with $d \geq 2$, it is the case that* $\theta(p_c) = 0$.

It is known that $\theta(p_c) = 0$ when either $d = 2$ (by results of [135], see Theorem 5.33) or $d \geq 19$ (by the lace expansion of [132, 133]). The claim is believed to be canonical of percolation models on all lattices and in all dimensions.

Suppose now that $p > p_c$, so that $\theta(p) > 0$. What can be said about the number N of infinite open clusters? Since the event $\{N \geq 1\}$ is translation-invariant, it is trivial under the product measure $\mathbb{P}_p$. However,

$$\mathbb{P}_p(N \geq 1) \geq \theta(p) > 0,$$

whence

$$\mathbb{P}_p(N \geq 1) = 1, \qquad p > p_c.$$

We shall see in the forthcoming Theorem 5.22 that $\mathbb{P}_p(N = 1) = 1$ whenever $\theta(p) > 0$, which is to say that there exists a unique infinite open cluster throughout the supercritical phase.

A supercritical percolation process in *two* dimensions may be studied in either of two ways. The first of these is by duality. Consider bond percolation on $\mathbb{L}^2$ with density p. The dual process (as in the proof of the upper bound of Theorem 3.2) is bond percolation with density $1 - p$. We shall see in Theorem 5.33 that the self-dual point $p = \frac{1}{2}$ is also the critical point. Thus, the dual of a supercritical process is subcritical, and this enables a study of supercritical percolation on $\mathbb{L}^2$. A similar argument is valid for certain other lattices, although the self-duality of the square lattice is special.

While duality is the technique for studying supercritical percolation in two dimensions, the process may also be studied by the block argument that follows. The block method was devised expressly for three and more dimensions in the hope that, amongst other things, it would imply the claim of Conjecture 5.16. Block arguments are a work-horse of the theory of general interacting systems.

We assume henceforth that $d \geq 3$ and that p is such that $\theta(p) > 0$; under this hypothesis, we wish to gain some control of the geometry of the infinite open paths. The main result is the following, of which an outline proof is included later in the section. Let $A \subseteq \mathbb{Z}^d$, and write $p_c(A)$ for the critical probability of bond percolation on the subgraph of $\mathbb{L}^d$ induced by A. Thus, for example, $p_c = p_c(\mathbb{Z}^d)$. Recall that $\Lambda(k) = [-k, k]^d$.

5.17 Theorem [115]. *Let $d \geq 3$. If F is an infinite connected subset of $\mathbb{Z}^d$ with $p_c(F) < 1$, then for each $\eta > 0$ there exists an integer k such that*

$$p_c(2kF + \Lambda(k)) \leq p_c + \eta.$$

That is, for any set F sufficiently large that $p_c(F) < 1$, we may 'fatten' F to a set having critical probability as close to p_c as required. One particular application of this theorem is to the limit of slab critical probabilities, and we elaborate on this next.

Many results have been proved for subcritical percolation under the 'finite susceptibility' hypothesis that $\chi(p) < \infty$. The validity of this hypothesis for $p < p_c$ is implied by Theorem 5.1. Similarly, several important results for supercritical percolation have been proved under the hypothesis that 'percolation occurs in slabs'. The two-dimensional slab F_k of thickness $2k$ is the set

$$F_k = \mathbb{Z}^2 \times [-k, k]^{d-2} = \left(\mathbb{Z}^2 \times \{0\}^{d-2}\right) + \Lambda(k),$$

with critical probability $p_c(F_k)$. Since $F_k \subseteq F_{k+1} \subseteq \mathbb{Z}^d$, the decreasing limit $p_c(F) = \lim_{k \to \infty} p_c(F_k)$ exists and satisfies $p_c(F) \geq p_c$. The hypothesis of 'percolation in slabs' is that $p > p_c(F)$. By Theorem 5.17,

(5.18) $$\lim_{k \to \infty} p_c(F_k) = p_c.$$

One of the best examples of the use of 'slab percolation' is the following estimate of the extent of a finite open cluster. It asserts the exponential decay of a 'truncated' connectivity function when $d \geq 3$. A similar result may be proved by duality for $d = 2$.

Figure 5.3. Images of the Wulff crystal in two dimensions. These are in fact images created by numerical simulation of the Ising model, but the general features are similar to those of percolation. The simulations were for finite time, and the images are therefore only approximations to the true crystals. The pictures are 1024 pixels square, and the Ising inverse-temperatures are $\beta = \frac{4}{3}, \frac{10}{11}$. The corresponding random-cluster models have $q = 2$ and $p = 1 - e^{-4/3}, 1 - e^{-10/11}$, so that the right-hand picture is closer to criticality than the left.

5.19 Theorem [65]. *Let $d \geq 3$. The limit*

$$\sigma(p) = \lim_{n \to \infty} \left\{ -\frac{1}{n} \log \mathbb{P}_p(0 \leftrightarrow \partial \Lambda(n), |C| < \infty) \right\}$$

exists. Furthermore $\sigma(p) > 0$ if $p > p_c$.

We turn briefly to a discussion of the so-called 'Wulff crystal', illustrated in Figure 5.3. Much attention has been paid to the sizes and shapes of clusters formed in models of statistical mechanics. When a cluster C is infinite with a strictly positive probability, but is constrained to have some large *finite* size n, then C is said to form a large 'droplet'. The asymptotic shape of such a droplet as $n \to \infty$ is prescribed in general terms by the theory of the so-called Wulff crystal, see the original paper [243] of Wulff. Specializing to percolation, we ask for properties of the open cluster C at the origin, conditioned on the event $\{|C| = n\}$.

The study of the Wulff crystal is bound up with the law of the volume of a finite cluster. This has a tail that is 'quenched exponential',

(5.20) $$\mathbb{P}_p(|C| = n) \approx \exp(-\rho n^{(d-1)/d}),$$

where $\rho = \rho(p) \in (0, \infty)$ for $p > p_c$, and $\approx$ is to be interpreted in terms of exponential asymptotics. The explanation for the curious exponent is as follows. The 'most economic' way to create a large finite cluster is to find a region R containing a connected component D of size n, satisfying $D \leftrightarrow \infty$, and then to cut all connections leaving R. Since $p > p_c$, such regions R exist with $|R|$ (respectively, $|\partial R|$) having order n (respectively, $n^{(d-1)/d}$), and the 'cost' of the construction is exponential in $|\partial R|$.

The above argument yields a lower bound for $\mathbb{P}_p(|C| = n)$ of quenched-exponential type, but considerably more work is required to show the exact asymptotic of (5.20), and indeed one obtains more. The (conditional) shape of $Cn^{-1/d}$ converges as $n \rightarrow \infty$ to the solution of a certain variational problem, and the asymptotic region is termed the 'Wulff crystal' for the model. This is not too hard to make rigorous when $d = 2$, since the external boundary of C is then a closed curve. Serious technical difficulties arise when pursuing this programme when $d \geq 3$. See [60] for an account and a bibliography.

Outline proof of Theorem 5.19. The existence of the limit is an exercise in subadditivity of a standard type, although with some complications in this case (see [64, 106]). We sketch here a proof of the important estimate $\sigma(p) > 0$.

Let S_k be the $(d-1)$-dimensional slab

$$S_k = [0, k] \times \mathbb{Z}^{d-1}.$$

Since $p > p_c$, we have by Theorem 5.17 that $p > p_c(S_k)$ for some k, and we choose k accordingly. Let H_n be the hyperplane of vertices x of $\mathbb{L}^d$ with $x_1 = n$. It suffices to prove that

(5.21) $$\mathbb{P}_p(0 \leftrightarrow H_n, \ |C| < \infty) \leq e^{-\gamma n}$$

for some $\gamma = \gamma(p) > 0$. Define the slabs

$$T_i = \{x \in \mathbb{Z}^d : (i-1)k \leq x_1 < ik\}, \qquad 1 \leq i < \lfloor n/k \rfloor.$$

Any path from 0 to H_n traverses each T_i. Since $p > p_c(S_k)$, each slab contains (almost surely) an infinite open cluster (see Figure 5.4). If $0 \leftrightarrow H_n$ and $|C| < \infty$, then all paths from 0 to H_n must evade all such clusters. There are $\lfloor n/k \rfloor$ slabs to traverse, and a price is paid for each. Modulo a touch of rigour, this implies that

$$\mathbb{P}_p(0 \leftrightarrow H_n, \ |C| < \infty) \leq [1 - \theta_k(p)]^{\lfloor n/k \rfloor},$$

where

$$\theta_k(p) = \mathbb{P}_p(0 \leftrightarrow \infty \text{ in } S_k) > 0.$$

The inequality $\sigma(p) > 0$ is proved. $\square$

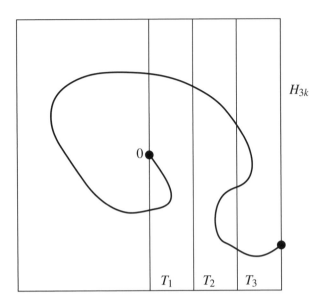

Figure 5.4. All paths from the origin to H_{3k} traverse the regions T_i, $i = 1, 2, 3$.

Outline proof of Theorem 5.17. The full proof can be found in [106, 115]. For simplicity, we take $F = \mathbb{Z}^2 \times \{0\}^{d-2}$, so that

$$2kF + \Lambda(k) = \mathbb{Z}^2 \times [-k, k]^{d-2}.$$

There are two main steps in the proof. In the first, we show the existence of long finite paths. In the second, we show how to take such finite paths and build an infinite cluster in a slab.

The principal parts of the first step are as follows. Let p be such that $\theta(p) > 0$.

1. Let $\epsilon > 0$. Since $\theta(p) > 0$, there exists m such that

$$\mathbb{P}_p(\Lambda(m) \leftrightarrow \infty) > 1 - \epsilon.$$

   [*This holds since there exists, almost surely, an infinite open cluster.*]

2. Let $n \geq 2m$, say, and let $k \geq 1$. We may choose n sufficiently large that, with probability at least $1 - 2\epsilon$, $\Lambda(m)$ is joined to at least k points in $\partial\Lambda(n)$. [*If, for some k, this fails for unbounded n, then there exists $N > m$ such that $\Lambda(m) \nleftrightarrow \partial\Lambda(N)$.*]

3. By choosing k sufficiently large, we may ensure that, with probability at least $1 - 3\epsilon$, $\Lambda(m)$ is joined to some point of $\partial\Lambda(n)$, which is itself connected to a copy of $\Lambda(m)$, lying 'on' the surface $\partial\Lambda(n)$ and

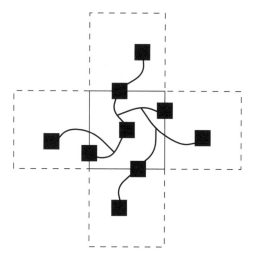

Figure 5.5. An illustration of the event that the block centred at the origin is open. Each black square is a seed.

every edge of which is open. [*We may choose k sufficiently large that there are many non-overlapping copies of* $\Lambda(m)$ *in the correct positions, indeed sufficiently many that, with high probability, one is totally open.*]

4. The open copy of $\Lambda(m)$, constructed above, may be used as a 'seed' for iterating the above construction. When doing this, we shall need some control over where the seed is placed. It may be shown that every face of $\partial\Lambda(n)$ contains (with large probability) a point adjacent to some seed, and indeed many such points. See Figure 5.5. [*There is sufficient symmetry to deduce this by the FKG inequality.*]

Above is the scheme for constructing long finite paths, and we turn to the second step.

5. This construction is now iterated. At each stage there is a certain (small) probability of failure. In order that there be a strictly positive probability of an infinite sequence of successes, we iterate in two 'independent' directions. With care, we may show that the construction dominates a certain supercritical site percolation process on $\mathbb{L}^2$.

6. We wish to deduce that an infinite sequence of successes entails an infinite open path of $\mathbb{L}^d$ within the corresponding slab. There are two difficulties with this. First, since we do not have total control of the positions of the seeds, the actual path in $\mathbb{L}^d$ may leave every slab. This may be overcome by a process of 'steering', in which, at each stage, we

choose a seed in such a position as to compensate for earlier deviations in space.

7. A greater problem is that, in iterating the construction, we carry with us a mixture of 'positive' and 'negative' information (of the form that 'certain paths exist' and 'others do not'). In combining events, we cannot use the FKG inequality. The practical difficulty is that, although we may have an infinite sequence of successes, there will generally be breaks in any corresponding open route to ∞. This is overcome by sprinkling down a few more open edges, that is, by working at edge-density $p + \delta$ where $\delta > 0$, rather than at density p.

In conclusion, we find that, if $\theta(p) > 0$ and $\delta > 0$, then there exists, with large probability, an infinite $(p + \delta)$-open path in a slab of the form $T_k = \mathbb{Z}^2 \times [-k, k]^{d-2}$ for sufficiently large k. The claim of the theorem follows.

There are many details to be considered in carrying out the above programme, and these are omitted here. $\qquad\square$

5.3 Uniqueness of the infinite cluster

The principal result of this section is the following: for any value of p for which $\theta(p) > 0$, there exists (almost surely) a unique infinite open cluster. Let $N = N(\omega)$ be the number of infinite open clusters.

5.22 Theorem [12]. *If $\theta(p) > 0$, then $\mathbb{P}_p(N = 1) = 1$.*

A similar conclusion holds for more general probability measures. The two principal ingredients of the generalization are the translation-invariance of the measure, and the so-called 'finite-energy property' that states that, conditional on the states of all edges except e, say, the state of e is 0 (respectively, 1) with a strictly positive (conditional) probability.

Proof. We follow [55]. The claim is trivial if $p = 0, 1$, and we assume henceforth that $0 < p < 1$. Let $S = S(n)$ be the 'diamond' $S(n) = \{x \in \mathbb{Z}^d : \delta(0, x) \le n\}$, and let $\mathbb{E}_S$ be the set of edges of $\mathbb{L}^d$ joining pairs of vertices in S. We write $N_S(0)$ (respectively, $N_S(1)$) for the total number of infinite open clusters when all edges in $\mathbb{E}_S$ are declared to be closed (respectively, open). Finally, M_S denotes the number of infinite open clusters that intersect S.

The sample space $\Omega = \{0, 1\}^{\mathbb{E}^d}$ is a product space with a natural family of translations, and $\mathbb{P}_p$ is a product measure on Ω. Since N is a translation-invariant function on Ω, it is almost surely constant, which is to say that

(5.23) $\exists k = k(p) \in \{0, 1, 2, \ldots\} \cup \{\infty\}$ such that $\mathbb{P}_p(N = k) = 1$.

Next we show that the k in (5.23) necessarily satisfies $k \in \{0, 1, \infty\}$. Suppose that (5.23) holds with $k < \infty$. Since every configuration on $\mathbb{E}_S$ has a strictly positive probability, it follows by the almost-sure constantness of N that

$$\mathbb{P}_p\big(N_S(0) = N_S(1) = k\big) = 1.$$

Now $N_S(0) = N_S(1)$ if and only if S intersects at most one infinite open cluster (this is where we use the assumption that $k < \infty$), and therefore

$$\mathbb{P}_p(M_S \geq 2) = 0.$$

Clearly, M_S is non-decreasing in $S = S(n)$, and $M_{S(n)} \to N$ as $n \to \infty$. Therefore,

(5.24) $$0 = \mathbb{P}_p(M_{S(n)} \geq 2) \to \mathbb{P}_p(N \geq 2),$$

which is to say that $k \leq 1$.

It remains to rule out the case $k = \infty$. Suppose that $k = \infty$. We will derive a contradiction by using a geometrical argument. We call a vertex x a *trifurcation* if:

(a) x lies in an infinite open cluster,

(b) there exist exactly three open edges incident to x, and

(c) the deletion of x and its three incident open edges splits this infinite cluster into exactly three disjoint infinite clusters and no finite clusters;

Let T_x be the event that x is a trifurcation. By translation-invariance, $\mathbb{P}_p(T_x)$ is constant for all x, and therefore

(5.25) $$\frac{1}{|S(n)|} \mathbb{E}_p\left(\sum_{x \in S(n)} 1_{T_x}\right) = \mathbb{P}_p(T_0).$$

It will be useful to know that the quantity $\mathbb{P}_p(T_0)$ is strictly positive, and it is here that we use the assumed infinity of infinite clusters. Let $M_S(0)$ be the number of infinite open clusters that intersect S when all edges of $\mathbb{E}_S$ are declared closed. Since $M_S(0) \geq M_S$, by the remarks around (5.24),

$$\mathbb{P}_p(M_{S(n)}(0) \geq 3) \geq \mathbb{P}_p(M_{S(n)} \geq 3) \to \mathbb{P}_p(N \geq 3) = 1 \qquad \text{as } n \to \infty.$$

Therefore, there exists m such that

$$\mathbb{P}_p(M_{S(m)}(0) \geq 3) \geq \tfrac{1}{2}.$$

We set $S = S(m)$ and $\partial S = S(m) \setminus S(m-1)$. Note that:

(a) the event $\{M_S(0) \geq 3\}$ is independent of the states of edges in $\mathbb{E}_S$,

(b) if the event $\{M_S(0) \geq 3\}$ occurs, there exist $x, y, z \in \partial S$ lying in distinct infinite open clusters of $\mathbb{E}^d \setminus \mathbb{E}_S$.

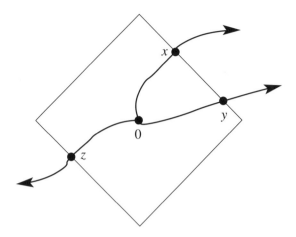

Figure 5.6.   Take a diamond S that intersects at least three distinct
infinite open clusters, and then alter the configuration inside S in order
to create a configuration in which 0 is a trifurcation.

Let $\omega \in \{M_S(0) \geq 3\}$, and pick $x = x(\omega)$, $y = y(\omega)$, $z = z(\omega)$
according to (b). If there is more than one possible such triple, we pick such
a triple according to some predetermined rule. It is a minor geometrical
exercise (see Figure 5.6) to verify that there exist in $\mathbb{E}_S$ three paths joining
the origin to (respectively) x, y, and z, and that these paths may be chosen
in such a way that:

 (i) the origin is the unique vertex common to any two of them, and

 (ii) each touches exactly one vertex lying in ∂S.

Let $J_{x,y,z}$ be the event that all the edges in these paths are open, and that all
other edges in $\mathbb{E}_S$ are closed.

Since S is finite,

$$\mathbb{P}_p(J_{x,y,z} \mid M_S(0) \geq 3) \geq \big[\min\{p, 1-p\}\big]^R > 0,$$

where $R = |\mathbb{E}_S|$. Now,

$$\mathbb{P}_p(0 \text{ is a trifurcation}) \geq \mathbb{P}_p(J_{x,y,z} \mid M_S(0) \geq 3)\mathbb{P}_p(M_S(0) \geq 3)$$

$$\geq \tfrac{1}{2}\big[\min\{p, 1-p\}\big]^R > 0,$$

which is to say that $\mathbb{P}_p(T_0) > 0$ in (5.25).

It follows from (5.25) that the mean number of trifurcations inside $S =$
$S(n)$ grows in the manner of $|S|$ as $n \to \infty$. On the other hand, we shall see
next that the number of trifurcations inside S can be no larger than the size of
the boundary of S, and this provides the necessary contradiction. This final

step must be performed properly (see [55, 106]), but the following rough argument is appealing and may be made rigorous. Select a trifurcation (t_1, say) of S, and choose some vertex $y_1 \in \partial S$ such that $t_1 \leftrightarrow y_1$ in S. We now select a new trifurcation $t_2 \in S$. It may be seen, using the definition of the term 'trifurcation', that there exists $y_2 \in \partial S$ such that $y_1 \neq y_2$ and $t_2 \leftrightarrow y_2$ in S. We continue similarly, at each stage picking a new trifurcation $t_k \in S$ and a new vertex $y_k \in \partial S$. If there are τ trifurcations in S, then we obtain τ distinct vertices y_k of ∂S. Therefore, $|\partial S| \geq \tau$. However, by the remarks above, $\mathbb{E}_p(\tau)$ is comparable to S. This is a contradiction for large n, since $|\partial S|$ grows in the manner of n^{d-1} and $|S|$ grows in the manner of n^d. □

5.4 Phase transition

Macroscopic functions, such as the percolation probability and mean cluster size,

$$\theta(p) = \mathbb{P}_p(|C| = \infty), \quad \chi(p) = \mathbb{E}_p|C|,$$

have singularities at $p = p_c$, and there is overwhelming evidence that these are of 'power law' type. A great deal of effort has been invested by physicists and mathematicians towards understanding the nature of the percolation phase-transition. The picture is now fairly clear when $d = 2$, owing to the very significant progress in recent years in relating critical percolation to the Schramm–Löwner curve SLE_6. There remain however substantial difficulties to be overcome before this chapter of percolation theory can be declared written, even when $d = 2$. The case of large d (currently, $d \geq 19$) is also well understood, through work based on the so-called 'lace expansion'. Most problems remain open in the obvious case $d = 3$, and ambitious and brave students are thus directed with caution.

The nature of the percolation singularity is supposed to be canonical, in that it is expected to have certain general features in common with phase transitions of other models of statistical mechanics. These features are sometimes referred to as 'scaling theory' and they relate to 'critical exponents'. There are two sets of critical exponents, arising firstly in the limit as $p \to p_c$, and secondly in the limit over increasing distances when $p = p_c$. We summarize the notation in Table 5.7.

The asymptotic relation $\approx$ should be interpreted loosely (perhaps via logarithmic asymptotics[1]). The radius of C is defined by

$$\mathrm{rad}(C) = \sup\{\|x\| : 0 \leftrightarrow x\},$$

[1]We say that $f(x)$ is logarithmically asymptotic to $g(x)$ as $x \to 0$ (respectively, $x \to \infty$) if $\log f(x)/\log g(x) \to 1$. This is often written as $f(x) \approx g(x)$.

Function		Behaviour	Exp.
percolation probability	$\theta(p) = \mathbb{P}_p(|C| = \infty)$	$\theta(p) \approx (p - p_c)^\beta$	β
truncated mean cluster size	$\chi^f(p) = \mathbb{E}_p(|C|1_{|C|<\infty})$	$\chi^f(p) \approx |p - p_c|^{-\gamma}$	γ
number of clusters per vertex	$\kappa(p) = \mathbb{E}_p(|C|^{-1})$	$\kappa'''(p) \approx |p - p_c|^{-1-\alpha}$	α
cluster moments	$\chi_k^f(p) = \mathbb{E}_p(|C|^k 1_{|C|<\infty})$	$\dfrac{\chi_{k+1}^f(p)}{\chi_k^f(p)} \approx |p - p_c|^{-\Delta}, \ k \geq 1$	Δ
correlation length	$\xi(p)$	$\xi(p) \approx |p - p_c|^{-\nu}$	ν
cluster volume		$\mathbb{P}_{p_c}(|C| = n) \approx n^{-1-1/\delta}$	δ
cluster radius		$\mathbb{P}_{p_c}(\text{rad}(C) = n) \approx n^{-1-1/\rho}$	ρ
connectivity function		$\mathbb{P}_{p_c}(0 \leftrightarrow x) \approx \|x\|^{2-d-\eta}$	η

Table 5.7. Eight functions and their critical exponents.

where

$$\|x\| = \sup_i |x_i|, \qquad x = (x_1, x_2, \dots, x_d) \in \mathbb{Z}^d,$$

is the supremum (L^∞) norm on $\mathbb{Z}^d$. The limit as $p \to p_c$ should be interpreted in a manner appropriate for the function in question (for example, as $p \downarrow p_c$ for $\theta(p)$, but as $p \to p_c$ for $\kappa(p)$).

There are eight critical exponents listed in Table 5.7, denoted α, β, γ, δ, ν, η, ρ, Δ, but there is no general proof of the existence of any of these exponents for arbitrary d. In general, the eight critical exponents may be defined for phase transitions in a quite large family of physical systems. However, it is not believed that they are independent variables, but rather that they satisfy the *scaling relations*

$$2 - \alpha = \gamma + 2\beta = \beta(\delta + 1),$$
$$\Delta = \delta\beta,$$
$$\gamma = \nu(2 - \eta),$$

and, when d is not too large, the *hyperscaling relations*

$$d\rho = \delta + 1,$$
$$2 - \alpha = d\nu.$$

The *upper critical dimension* is the largest value d_c such that the hyperscaling relations hold for $d \le d_c$. It is believed that $d_c = 6$ for percolation. There is no general proof of the validity of the scaling and hyperscaling relations, although quite a lot is known when $d = 2$ and for large d.

In the context of percolation, there is an analytical rationale behind the scaling relations, namely the 'scaling hypotheses' that

$$\mathbb{P}_p(|C| = n) \sim n^{-\sigma} f\left(n/\xi(p)^\tau\right)$$
$$\mathbb{P}_p(0 \leftrightarrow x, \ |C| < \infty) \sim \|x\|^{2-d-\eta} g\left(\|x\|/\xi(p)\right)$$

in the double limit as $p \to p_c$, $n \to \infty$, and for some constants σ, τ, η and functions f, g. Playing loose with rigorous mathematics, the scaling relations may be derived from these hypotheses. Similarly, the hyperscaling relations may be shown to be not too unreasonable, at least when d is not too large. For further discussion, see [106].

We note some further points.

Universality. It is believed that the numerical values of critical exponents depend only on the value of d, and are independent of the particular percolation model.

Two dimensions. When $d = 2$, perhaps

$$\alpha = -\tfrac{2}{3}, \ \beta = \tfrac{5}{36}, \ \gamma = \tfrac{43}{18}, \ \delta = \tfrac{91}{5}, \dots$$

See (5.45).

Large dimension. When d is sufficiently large (actually, $d \ge d_c$) it is believed that the critical exponents are the same as those for percolation on a tree (the 'mean-field model'), namely $\delta = 2$, $\gamma = 1$, $\nu = \tfrac{1}{2}$, $\rho = \tfrac{1}{2}$, and so on (the other exponents are found to satisfy the scaling relations). Using the first hyperscaling relation, this is consistent with the contention that $d_c = 6$. Such statements are known to hold for $d \ge 19$, see [132, 133] and the remarks later in this section.

Open challenges include to prove:
- the existence of critical exponents,
- universality,
- the scaling and hyperscaling relations,
- the conjectured values when $d = 2$,
- the conjectured values when $d \geq 6$.

Progress towards these goals has been positive. For sufficiently large d, exact values are known for many exponents, namely the values from percolation on a regular tree. There has been remarkable progress in recent years when $d = 2$, inspired largely by work of Schramm [215], enacted by Smirnov [222], and confirmed by the programme pursued by Lawler, Schramm, and Werner to understand SLE curves. See Section 5.6.

We close this section with some further remarks on the case of large d. The expression 'mean-field' permits several interpretations depending on context. A narrow interpretation of the term 'mean-field theory' for percolation involves trees rather than lattices. For percolation on a regular tree, it is quite easy to perform exact calculations of many quantities, including the numerical values of critical exponents. That is, $\delta = 2$, $\gamma = 1$, $\nu = \frac{1}{2}$, $\rho = \frac{1}{2}$, and other exponents are given according to the scaling relations, see [106, Chap. 10].

Turning to percolation on $\mathbb{L}^d$, it is known as remarked above that the critical exponents agree with those of a regular tree when d is sufficiently large. In fact, this is believed to hold if and only if $d \geq 6$, but progress so far assumes that $d \geq 19$. In the following theorem, we write $f(x) \simeq g(x)$ if there exist positive constants c_1, c_2 such that $c_1 f(x) \leq g(x) \leq c_2 f(x)$ for all x close to a limiting value.

5.26 Theorem [133]. *For $d \geq 19$,*

$$\theta(p) \simeq (p - p_c)^1 \qquad \text{as } p \downarrow p_c,$$

$$\chi(p) \simeq (p_c - p)^{-1} \qquad \text{as } p \uparrow p_c,$$

$$\xi(p) \simeq (p_c - p)^{-\frac{1}{2}} \qquad \text{as } p \uparrow p_c,$$

$$\frac{\chi^{\mathrm{f}}_{k+1}(p)}{\chi^{\mathrm{f}}_k(p)} \simeq (p_c - p)^{-2} \qquad \text{as } p \uparrow p_c, \text{for } k \geq 1.$$

Note the strong form of the asymptotic relation $\simeq$, and the identification of the critical exponents β, γ, Δ, ν. The proof of Theorem 5.26 centres on a property known as the 'triangle condition'. Define

$$(5.27) \qquad T(p) = \sum_{x,y \in \mathbb{Z}^d} \mathbb{P}_p(0 \leftrightarrow x)\mathbb{P}_p(x \leftrightarrow y)\mathbb{P}_p(y \leftrightarrow 0),$$

and consider the *triangle condition*

$$T(p_c) < \infty.$$

The triangle condition was introduced by Aizenman and Newman [15], who showed that it implied that $\chi(p) \simeq (p_c - p)^{-1}$ as $p \uparrow p_c$. Subsequently other authors showed that the triangle condition implied similar asymptotics for other quantities. It was Takashi Hara and Gordon Slade [132] who verified the triangle condition for large d, exploiting a technique known as the 'lace expansion'.

5.5 Open paths in annuli

The remainder of this chapter is devoted to percolation in two dimensions, in the context of either the site model on the triangular lattice $\mathbb{T}$ or the bond model on the square lattice $\mathbb{L}^2$.

There is a very useful technique for building open paths with certain geometry in two dimensions. It leads to a proof that the chance of an open cycle within an annulus $[-3n, 3n]^2 \setminus [-n, n]^2$ is at least $f(\delta)$, where δ is the chance of an open crossing of the square $[-n, n]^2$, and f is a strictly positive function. This result was useful in some of the original proofs concerning the critical probability of bond percolation on $\mathbb{L}^2$ (see [106, Sect. 11.7]), and it has re-emerged more recently as central to estimates that permit the proof of Cardy's formula and conformal invariance. It is commonly named after Russo [211] and Seymour–Welsh [221]. The RSW lemma will be stated and proved in this section, and utilized in the next three. Since our application in Sections 5.6–5.7 is to site percolation on the triangular lattice, we shall phrase the RSW lemma in that context. It is left to the reader to adapt and develop the arguments of this section for bond percolation on the square lattice (see Exercise 5.5). The triangular lattice $\mathbb{T}$ is drawn in Figure 5.8, together with its dual hexagonal lattice $\mathbb{H}$.

There is a special property in common to the bond model on $\mathbb{L}^2$ and the site model on $\mathbb{T}$, namely that the 'external' boundary of a finite open cluster contains a closed cycle. This was illustrated in Figure 3.1 for bond percolation on $\mathbb{L}^2$, and may be seen similarly for $\mathbb{T}$. This property is central to the proofs that these models have critical probability $p_c = \frac{1}{2}$.

RSW theory is presented in [106, Sect. 11.7] for the square lattice $\mathbb{L}^2$ and general bond-density p. We could follow the same route here for the triangular lattice, but for the sake of variation (and with an eye to later applications) we shall restrict ourselves to the case $p = \frac{1}{2}$ and shall give a shortened proof due to Stanislav Smirnov. The more conventional approach may be found in [238], see also [237], and [46] for a variant on the square

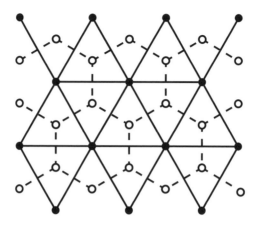

Figure 5.8. The triangular lattice $\mathbb{T}$ and the (dual) hexagonal lattice $\mathbb{H}$.

lattice. Thus, in this section we restrict ourselves to site percolation on $\mathbb{T}$ with density $\frac{1}{2}$. Each site of $\mathbb{T}$ is coloured *black* with probability $\frac{1}{2}$, and *white* otherwise, and the relevant probability measure is denoted as $\mathbb{P}$.

The triangular lattice is embedded in $\mathbb{R}^2$ with vertex-set $\{m\mathbf{i} + n\mathbf{j} :$ $(m, n) \in \mathbb{Z}^2\}$, where $\mathbf{i} = (1, 0)$ and $\mathbf{j} = \frac{1}{2}(1, \sqrt{3})$. Write $R_{a,b}$ for the subgraph induced by vertices in the rectangle $[0, a] \times [0, b]$, and we shall restrict ourselves always to integers a and integer multiples b of $\frac{1}{2}\sqrt{3}$. We shall consider left–right crossings of rectangles R, and to this end we let its *left edge* $\mathrm{L}(R)$ (respectively, *right edge* $\mathrm{R}(R)$) be the set of vertices of R within distance $\frac{1}{2}$ of its left side (respectively, right side). This minor geometrical complication arises because the vertical lines of $\mathbb{L}^2$ are not connected subgraphs of $\mathbb{T}$. Let $H_{a,b}$ be the event that there exists a black path that traverses $R_{a,b}$ from $\mathrm{L}(R_{a,b})$ to $\mathrm{R}(R_{a,b})$. The 'engine room' of the RSW method is the following lemma.

5.28 Lemma. $\mathbb{P}(H_{2a,b}) \geq \frac{1}{4}\mathbb{P}(H_{a,b})^2.$

By iteration,

$$(5.29) \qquad \mathbb{P}(H_{2^k a,b}) \geq 4\big[\tfrac{1}{4}\mathbb{P}(H_{a,b})\big]^{2^k}, \qquad k \geq 0.$$

As 'input' to this inequality, we prove the following.

5.30 Lemma. *We have that* $\mathbb{P}(H_{a,a\sqrt{3}}) \geq \frac{1}{2}.$

Let Λ_m be the set of vertices in $\mathbb{T}$ at graph-theoretic distance m or less from the origin 0, and define the annulus $A_n = \Lambda_{3n} \setminus \Lambda_{n-1}$. Let O_n be the event that A_n contains a black cycle C such that 0 lies in the bounded component of $\mathbb{R}^2 \setminus C$.

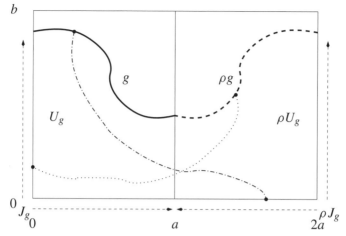

Figure 5.9.   The crossing g and its reflection ρg in the box $R_{2a,b}$. The events B_g and $W_{\rho g}$ are illustrated by the two lower paths, and exactly one of these events occurs.

5.31 Theorem (RSW). *There exists $\sigma > 0$ such that $\mathbb{P}(O_n) > \sigma$ for all $n \geq 1$.*

Proof of Lemma 5.28. We follow an unpublished argument of Stanislav Smirnov.[2] Let g be a path that traverses $R_{a,b}$ from left to right. Let ρ denote reflection in the line $x = a$, so that ρg connects the left and right edges of $[a, 2a] \times [0, b]$. See Figure 5.9. Assume for the moment that g does not intersect the x-axis. Let U_g be the connected subgraph of $R_{a,b}$ lying 'strictly beneath' g, and $\overline{U_g}$ the corresponding graph lying 'on or beneath' g. Let J_g (respectively, J_b) be the part of the boundary ∂U_g (respectively, $R_{a,b}$) lying on either the x-axis or y-axis but not in g, and let ρJ_g (respectively, ρJ_b) be its reflection.

Next we use the self-duality of site percolation on $\mathbb{T}$. Let B_g be the event that there exists a path of $\overline{U_g} \cup \rho U_g$ joining some vertex of g to some vertex of ρJ_g, with the property that every vertex not belonging to g is black. Let $W_{\rho g}$ be defined similarly in terms of a white path of $U_g \cup \rho \overline{U_g}$ from ρg to J_g. The key fact is the following: $W_{\rho g}$ occurs whenever B_g does not. This holds as follows. Assume B_g does not occur. The set of vertices reached along black paths from g does not intersect ρJ_g. Its external boundary (away from $g \cup \rho g$) is white and connected, and thus contains a path of the sort required for $W_{\rho g}$. There is a complication that does not arise for the bond model on $\mathbb{L}^2$, namely that *both* B_g and $W_{\rho g}$ can occur if the right endvertex of g lies

---

[2]See also [236].

on the line $x = a$.

By symmetry, $\mathbb{P}(B_g) = \mathbb{P}(W_{\rho g})$, and by the above,

(5.32) $$\mathbb{P}(B_g) = \mathbb{P}(W_{\rho g}) \geq \tfrac{1}{2}.$$

The same holds if g touches the x-axis, with J_g suitably adapted.

Let L be the left edge of $R_{2a,b}$ and R its right edge. By the FKG inequality,

$$\mathbb{P}(H_{2a,b}) \geq \mathbb{P}(\text{L} \leftrightarrow \rho J_b, \text{ R} \leftrightarrow J_b)$$

$$\geq \mathbb{P}(\text{L} \leftrightarrow \rho J_b)\mathbb{P}(\text{R} \leftrightarrow J_b) = \mathbb{P}(\text{L} \leftrightarrow \rho J_b)^2,$$

where $\leftrightarrow$ denotes connection by a black path.

Let γ be the 'highest' black path from the left to the right sides of $R_{a,b}$, if such a path exists. Conditional on the event $\{\gamma = g\}$, the states of sites beneath g are independent Bernoulli variables, whence, in particular, the events B_g and $\{\gamma = g\}$ are independent. Therefore,

$$\mathbb{P}(\text{L} \leftrightarrow \rho J_b) = \sum_g \mathbb{P}(\gamma = g, \ B_g) = \sum_g \mathbb{P}(B_g)\mathbb{P}(\gamma = g)$$

$$\geq \tfrac{1}{2} \sum_g \mathbb{P}(\gamma = g) = \tfrac{1}{2}\mathbb{P}(H_{a,b})$$

by (5.32), and the lemma is proved.   □

Proof of Lemma 5.30. This is similar to the argument leading to (5.32). Consider the rhombus R of $\mathbb{T}$ comprising all vertices of the form $m\mathbf{i} + n\mathbf{j}$ for $0 \leq m, n \leq 2a$. Let B be the event that R is traversed from left to right by a black path, and W the event that it is traversed from top to bottom by a white path. These two events are mutually exclusive with the same probability, and one or the other necessarily occurs. Therefore, $\mathbb{P}(B) = \tfrac{1}{2}$. On B, there exists a left–right crossing of the (sub-)rectangle $[a, 2a] \times [0, a\sqrt{3}]$, and the claim follows.   □

Proof of Theorem 5.31. By (5.29) and Lemma 5.30, there exists $\alpha > 0$ such that

$$\mathbb{P}(H_{8n,n\sqrt{3}}) \geq \alpha, \qquad n \geq 1.$$

We may represent the annulus A_n as the pairwise-intersection of six copies of $R_{8n,n\sqrt{3}}$ obtained by translation and rotation, illustrated in Figure 5.10. If each of these is traversed by a black path in its long direction, then the event O_n occurs. By the FKG inequality,

$$\mathbb{P}(O_n) \geq \alpha^6,$$

and the theorem is proved.   □

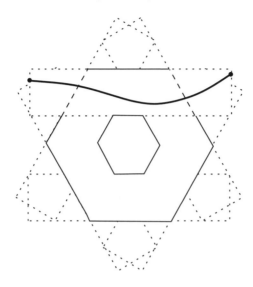

Figure 5.10. If each of six long rectangles are traversed in the long direction by black paths, then the intersection of these paths contains a black cycle within the annulus A_n.

5.6 The critical probability in two dimensions

We revert to bond percolation on the square lattice in this section. The square lattice has a property of self-duality, illustrated in Figure 1.5. 'Percolation of open edges on the primal lattice' is dual to 'percolation of closed edges on the dual lattice'. The self-dual value of p is thus $p = \frac{1}{2}$, and it was long believed that the self-dual point is also the critical point p_c. Theodore Harris [135] proved by a geometric construction that $\theta(\frac{1}{2}) = 0$, whence $p_c(\mathbb{Z}^2) \geq \frac{1}{2}$. Harry Kesten [151] proved the complementary inequality.

5.33 Theorem [135, 151]. *The critical probability of bond percolation on the square lattice equals $\frac{1}{2}$. Furthermore, $\theta(\frac{1}{2}) = 0$.*

Before giving a proof, we make some comments on the original proof. Harris [135] showed that, if $\theta(\frac{1}{2}) > 0$, then we can construct closed dual cycles around the origin. Such cycles prevent the cluster C from being infinite, and therefore $\theta(\frac{1}{2}) = 0$, a contradiction. Similar 'path-construction' arguments were developed by Russo [211] and Seymour–Welsh [221] in a proof that $p > p_c$ if and only if $\chi(1 - p) < \infty$. This so-called 'RSW method' has acquired prominence through recent work on SLE (see Sections 5.5 and 5.7).

The complementary inequality $p_c(\mathbb{Z}^2) \leq \frac{1}{2}$ was proved by Kesten in

[151]. More specifically, he showed that, for $p < \frac{1}{2}$, the probability of an open left–right crossing of the rectangle $[0, 2^k] \times [0, 2^{k+1}]$ tends to zero as $k \to \infty$. With the benefit of hindsight, we may view his argument as establishing a type of sharp-threshold theorem for the event in question.

The arguments that prove Theorem 5.33 may be adapted to certain other situations. For example, Wierman [238] has proved that the critical probabilities of bond percolation on the hexagonal/triangular pair of lattices (see Figure 5.8) are the dual pair of values satisfying the star–triangle transformation. Russo [212] adapted the arguments to site percolation on the square lattice. It is easily seen by the same arguments that site percolation on the triangular lattice has critical probability $\frac{1}{2}$.[3]

The proof of Theorem 5.33 is broken into two parts.

Proof of Theorem 5.33: $\theta(\frac{1}{2}) = 0$, *and hence* $p_c \geq \frac{1}{2}$. Zhang discovered a beautiful proof of this, using only the uniqueness of the infinite cluster, see [106, Sect. 11.3]. Set $p = \frac{1}{2}$, and assume that $\theta(\frac{1}{2}) > 0$. Let $T(n) = [0, n]^2$, and find N sufficiently large that

$$\mathbb{P}_{\frac{1}{2}}(\partial T(n) \leftrightarrow \infty) > 1 - (\tfrac{1}{8})^4, \qquad n \geq N.$$

We set $n = N + 1$. Let A^l, A^r, A^t, A^b be the (respective) events that the left, right, top, bottom sides of $T(n)$ are joined to ∞ off $T(n)$. By the FKG inequality,

$$\mathbb{P}_{\frac{1}{2}}(T(n) \nleftrightarrow \infty) = \mathbb{P}_{\frac{1}{2}}(\overline{A^l} \cap \overline{A^r} \cap \overline{A^t} \cap \overline{A^b})$$
$$\geq \mathbb{P}_{\frac{1}{2}}(\overline{A^l}) P(\overline{A^r}) P(\overline{A^t}) P(\overline{A^b})$$
$$= \mathbb{P}_{\frac{1}{2}}(\overline{A^g})^4$$

by symmetry, for $g = l, r, t, b$. Therefore,

$$\mathbb{P}_{\frac{1}{2}}(A^g) \geq 1 - \mathbb{P}_{\frac{1}{2}}(T(n) \nleftrightarrow \infty)^{1/4} > \tfrac{7}{8}.$$

We consider next the dual box, with vertex set $T(n)_d = [0, n-1]^2 + (\frac{1}{2}, \frac{1}{2})$. Let A^l_d, A^r_d, A^t_d, A^b_d denote the (respective) events that the left, right, top, bottom sides of $T(n)_d$ are joined to ∞ by a *closed* dual path off $T(n)_d$. Since each edge of the dual is closed with probability $\frac{1}{2}$,

$$\mathbb{P}_{\frac{1}{2}}(A^g_d) > \tfrac{7}{8}, \qquad g = l, r, t, b.$$

Consider the event $A = A^l \cap A^r \cap A^t_d \cap A^b_d$, illustrated in Figure 5.11. Clearly, $\mathbb{P}_{\frac{1}{2}}(\overline{A}) \leq \frac{1}{2}$, so that $\mathbb{P}_{\frac{1}{2}}(A) \geq \frac{1}{2}$. However, on A, either $\mathbb{L}^2$ has

---

[3]See also Section 5.8.

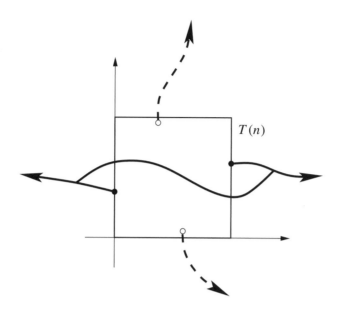

Figure 5.11. The left and right sides of the box $T(n)$ are joined to infinity by open paths of the primal lattice, and the top and bottom sides of the dual box $T(n)_\mathrm{d}$ are joined to infinity by closed dual paths. Using the uniqueness of the infinite open cluster, the two open paths must be joined by an open path. This forces the existence of two disjoint infinite closed clusters in the dual.

two infinite open clusters, or its dual has two infinite closed clusters. By Theorem 5.22, each event has probability 0, a contradiction. We deduce that $\theta(\frac{1}{2}) = 0$, implying in particular that $p_\mathrm{c} \geq \frac{1}{2}$. □

Proof of Theorem 5.33: $p_\mathrm{c} \leq \frac{1}{2}$. We give two proofs. The first uses the general exponential-decay Theorem 5.1. The second was proposed by Stanislav Smirnov, and avoids the appeal to Theorem 5.1. It is close in spirit to Kesten's original proof, and resonates with Menshikov's proof of Theorem 5.1. A third approach to the proof uses the sharp-threshold Theorem 4.81, and this is deferred to Section 5.8.

Proof A. Suppose instead that $p_\mathrm{c} > \frac{1}{2}$. By Theorem 5.1, there exists $\gamma > 0$ such that

(5.34) $$\mathbb{P}_{\frac{1}{2}}(0 \leftrightarrow \partial[-n, n]^2) \leq e^{-\gamma n}, \qquad n \geq 1.$$

Let $S(n)$ be the graph with vertex set $[0, n + 1] \times [0, n]$ and edge set containing all edges inherited from $\mathbb{L}^2$ except those in either the left side or the

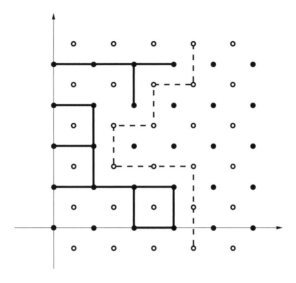

Figure 5.12. If there is no open left–right crossing of $S(n)$, there must exist a closed top–bottom crossing in the dual.

right side of $S(n)$. Let A be the event that there exists an open path joining the left side and right side of $S(n)$. If A does not occur, then the top side of the dual of $S(n)$ is joined to the bottom side by a closed dual path. Since the dual of $S(n)$ is isomorphic to $S(n)$, and since $p = \frac{1}{2}$, it follows that $\mathbb{P}_{\frac{1}{2}}(A) = \frac{1}{2}$ (see Figure 5.12). However, by (5.34),

$$\mathbb{P}_{\frac{1}{2}}(A) \le (n+1)e^{-\gamma n},$$

a contradiction for large n. We deduce that $p_c \le \frac{1}{2}$.

Proof B. Let $\Lambda(k) = [-k, k]^2$, and let $A_k = \Lambda(3k) \setminus \Lambda(k)$ be an 'annulus'. The principal ingredient is an estimate that follows from the square-lattice version of the RSW Theorem 5.31.[4] Let $p = \frac{1}{2}$. There exist $c, \sigma > 0$ such that:

(a) there are at least $c \log r$ disjoint annuli A_k within $[-r, r]^2$,
(b) each such annulus contains, with probability at least α, a dual closed cycle having 0 in its inside.

Therefore, $g(r) = \mathbb{P}_{\frac{1}{2}}(0 \leftrightarrow \partial \Lambda(r))$ satisfies

$$(5.35) \qquad\qquad g(r) \le (1 - \sigma)^{c \log r} = r^{-\alpha},$$

[4]See Exercise 5.5.

where $\alpha = \alpha(c, \sigma) > 0$. For future use, let D be a random variable with

(5.36) $$\mathbb{P}(D \geq r) = g(r), \qquad r \geq 0.$$

There are a variety of ways of implementing the basic argument of this proof, of which we choose the following. Let $R_n = [0, 2n] \times [0, n]$, where $n \geq 1$, and let H_n be the event that R_n is traversed by an open path from left to right. The event A given in Proof A satisfies $\mathbb{P}_{\frac{1}{2}}(A) = \frac{1}{2}$. Hence, by Lemma 5.28 rewritten for the square lattice, there exists $\gamma > 0$ such that

(5.37) $$\mathbb{P}_{\frac{1}{2}}(H_n) \geq \gamma, \qquad n \geq 1.$$

This inequality will be used later in the proof.

We take $p \geq \frac{1}{2}$ and work with the dual model. Let S_n be the dual box $(\frac{1}{2}, \frac{1}{2}) + [0, 2n - 1] \times [0, n + 1]$, and let V_n be the event that S_n is traversed from top to bottom by a closed dual path. Let $N = N_n$ be the number of pivotal edges for the event V_n, and let Π be the event that $N \geq 1$ and all pivotal edges are closed (in the dual). We shall prove that

(5.38) $$\mathbb{E}_p(N \mid \Pi) \geq c'n^\alpha,$$

for some absolute positive constant c'.

For any top–bottom path λ of S_n, we write $L(\lambda)$ (respectively, $R(\lambda)$) for the set of edges of S_n lying strictly to the 'left' (respectively, 'right') of λ. On Π, there exists a closed top–bottom path of S_n, and from amongst such paths we may pick the leftmost, denoted Λ. As in the proof of Lemma 5.28, Λ is measurable on the states of edges in and to the left of Λ; that is to say, for any admissible path λ, the event $\{\Lambda = \lambda\}$ depends only on the states of edges in $\lambda \cup L(\lambda)$. (See Figure 5.13.)

Assume as above that Π occurs, and that $\Lambda = \lambda$. Every pivotal edge for V_n lies in λ. Each edge $e = \langle x, y \rangle \in \lambda$ has a dual edge $e_d = \langle u_l, u_r \rangle$, for some $u_l, u_r \in \mathbb{Z}^2$. Since λ is leftmost, exactly one of these endvertices, u_l say, is necessarily connected to the left side of R_n by an open primal path of edges dual to edges of $L(\lambda)$. In addition, e is pivotal for V_n if and only if u_r is connected to the right side of R_n by an open primal path.

We now take a walk along λ from bottom to top, encountering its edges in order. Let $f_1, f_2, \ldots, f_N$ be the pivotal edges thus encountered, with $f_i = \langle x_i, y_i \rangle$, and let y_0 be the initial vertex of λ and x_{N+1} the final vertex. Given Π, we have that $N \geq 1$, and there is a 'lowest' open path Ψ connecting the right side of R_n to an endvertex of the dual edge of f_1. By symmetry,

(5.39) $$\mathbb{P}_p(y_1 \text{ lies in the lower half of } S_n \mid \Pi) \geq \frac{1}{2}.$$

Consider now a configuration $\omega \in \Pi$, with $\Lambda = \lambda$ and $\Psi = \psi$, say. The states of edges in the region $T(\lambda, \psi)$ of S_n, lying both to the right of λ and

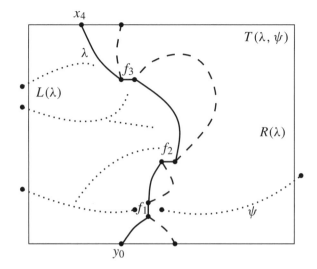

Figure 5.13. The leftmost closed top–bottom crossing λ. Primal vertices just to the 'left' of λ are connected by open (dotted) paths to the
left side of the rectangle. An edge f_1 of λ is pivotal if the vertex just
to its 'right' is joined by an open path to the right side. Between any
two successive pivotal edges, there exists a closed path lying entirely in
$R(\lambda)$. There are three pivotal edges f_i in this illustration, and the dashed
lines are the closed connections of $R(\lambda)$ joining successive f_i.

above ψ, are unaffected by the conditioning. That is, for given λ, ψ, the
states of the edges in $T(\lambda, \psi)$ are governed by product measure. What is
the distance from $f_1 = \langle x_1, y_1 \rangle$ to the next pivotal edge $f_2 = \langle x_2, y_2 \rangle$? No
pivotal edge in encountered on the way, and therefore there exists a closed
path of $T(\lambda, \psi)$ from y_1 to x_2. Since $1 - p \le \frac{1}{2}$, the L^∞ displacement of
such a path is no larger in distribution than a random variable D satisfying
(5.36). Having reached f_2, we iterate the argument until we attain the top
of S_n. The vertical displacement between two consecutive pivotal edges
is (conditional on the construction prior to the earlier such edge) bounded
above in distribution by $D + 1$, where the extra 1 takes care of the length of
an edge.

By (5.39) and the stochastic inequality,

(5.40) $\mathbb{P}_p(N \ge k + 1 \mid \Pi) \ge \frac{1}{2}\mathbb{P}(\Sigma_k \le n/2), \qquad k \ge 0,$

where $D' = 1 + \min\{D, n/2\}$ and $\Sigma_r = D'_1 + D'_2 + \cdots + D'_r$. Cf. (5.13).

There are at least two ways to continue. The first is to deduce that

$$2\mathbb{P}_p(N \geq k+1 \mid \Pi) \geq 1 - \mathbb{P}\big(D_i' > n/(2k) \text{ for some } 1 \leq i \leq k\big)$$
$$\geq 1 - kg\big(n/(2k) - 1\big).$$

We choose here to use instead the renewal theorem as in the proof of Theorem 5.1. Sum (5.40) over k to obtain as in (5.14) that

(5.41) $$\mathbb{E}_p(N \mid \Pi) \geq \tfrac{1}{2}\mathbb{E}(K),$$

where $K = \min\{k : \Sigma_k > n/2\}$. By Wald's equation,

$$\tfrac{1}{2}n < \mathbb{E}(\Sigma_K) = \mathbb{E}(K)\mathbb{E}(D'),$$

so that

(5.42) $$\mathbb{E}(K) > \frac{n/2}{\mathbb{E}(D')} = \frac{n/2}{\sum_{r=0}^{n/2} g(r)}.$$

Inequality (5.38) follows from (5.41)–(5.42) and (5.35)–(5.36).[5]

We prove next that

(5.43) $$\mathbb{P}_p(\Pi) \geq \mathbb{P}_p(H_n)\mathbb{P}_p(V_n).$$

Suppose V_n occurs, with $\Lambda = \lambda$, and let W_λ be the event that there exists $e \in \lambda$ such that: its dual edge $e_\mathrm{d} = \langle u, v \rangle$ has an endpoint connected to the right side of R_n by an open primal path of edges of $R(\lambda)$. The states of edges of $R(\lambda)$ are governed by product measure, so that

$$\mathbb{P}_p(W_l \mid \Lambda = \lambda) \geq \mathbb{P}_p(H_n).$$

Therefore,

$$\mathbb{P}_p(H_n)\mathbb{P}_p(V_n) \leq \sum_l \mathbb{P}_p(W_l \mid \Lambda = \lambda)\mathbb{P}_p(\Lambda = \lambda) = \mathbb{P}_p(\Pi).$$

Since $\mathbb{P}_p(H_n) = 1 - \mathbb{P}_p(V_n)$, by (5.38) and Russo's formula (Theorem 4.79),

$$\frac{d}{dp}\mathbb{P}_p(H_n) \geq \mathbb{E}_p(N) \geq c'n^\alpha \mathbb{P}_p(H_n)[1 - \mathbb{P}_p(H_n)], \qquad p \geq \tfrac{1}{2}.$$

The resulting differential inequality

$$\left[\frac{1}{\mathbb{P}_p(H_n)} + \frac{1}{1 - \mathbb{P}_p(H_n)}\right]\frac{d}{dp}\mathbb{P}_p(H_n) \geq c'n^\alpha$$

[5]Readers are invited to complete the details of the above argument.

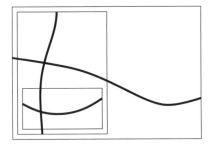

Figure 5.14. The boxes with aspect ratio 2 are arranged in such a way that, if all but finitely many are traversed in the long direction, then there must exist an infinite cluster.

may be integrated over the interval $[\frac{1}{2}, p]$ to obtain[6] via (5.37) that

$$1 - \mathbb{P}_p(H_n) \le \frac{1}{\gamma}\exp\{-c'(p - \tfrac{1}{2})n^\alpha\}.$$

This may be used to prove exponential decay in two dimensions (as in Theorems 5.1 and 5.3), but here we use only the (lesser) consequence that

(5.44) $$\sum_{n=1}^{\infty}[1 - \mathbb{P}_p(H_n)] < \infty, \qquad p > \tfrac{1}{2}.$$

We now use a block argument that was published in [63].[7] Let $p > \frac{1}{2}$. Consider the nested rectangles

$$B_{2r-1} = [0, 2^{2r}] \times [0, 2^{2r-1}], \quad B_{2r} = [0, 2^{2r}] \times [0, 2^{2r+1}], \qquad r \ge 1,$$

illustrated in Figure 5.14. Let K_{2r-1} (respectively, K_{2r}) be the event that B_{2r-1} (respectively, B_{2r}) is traversed from left to right (respectively, top to bottom) by an open path, so that $\mathbb{P}_p(K_k) = \mathbb{P}_p(H_{2^{k-1}})$. By (5.44) and the Borel–Cantelli lemma, all but finitely many of the K_k occur, $\mathbb{P}_p$-almost surely. By Figure 5.14 again, this entails the existence of an infinite open cluster, whence $\theta(p) > 0$, and hence $p_c \le \frac{1}{2}$. $\qquad\square$

5.7 Cardy's formula

There is a rich physical theory of phase transitions in theoretical physics, and critical percolation is at the heart of this theory. The case of two dimensions

---

[6]The same point may be reached using the theory of influence, as in Exercise 5.4.
[7]An alternative block argument may be found in Section 5.8.

is very special, in that methods of conformality and complex analysis, bolstered by predictions of conformal field theory, have given rise to a beautiful and universal vision for the nature of such singularities. This vision is both analytical and geometrical. Its proof has been one of the principal targets of probability theory and theoretical physics over recent decades. The 'road map' to the proof is now widely accepted, and many key ingredients have become clear. There remain some significant problems.

The principal ingredient of the mathematical theory is the SLE process introduced in Section 2.5. In a classical theorem of Löwner [177], we see that a growing path γ in $\mathbb{R}^2$ may be encoded via conformal maps g_t in terms of a so-called 'driving function' $b : [0, \infty) \rightarrow \mathbb{R}$. Oded Schramm [215] predicted that a variety of scaling limits of stochastic processes in $\mathbb{R}^2$ may be formulated thus, with b chosen as a Brownian motion with an appropriately chosen variance parameter κ. He gave a partial proof that LERW on $\mathbb{L}^2$, suitably re-scaled, has limit SLE_2, and he indicated that UST has limit SLE_8 and percolation SLE_6.

These observations did not come out of the blue. There was considerable earlier speculation around the idea of conformality, and we highlight the statement by John Cardy of his formula [59], and the discussions of Michael Aizenman and others concerning possible invariance under conformal maps (see, for example, [4, 5, 158]).

Much has been achieved since Schramm's paper [215]. Stanislav Smirnov [222, 223] has proved that critical site percolation on the triangular lattice satisfies Cardy's formula, and his route to 'complete conformality' and SLE_6 has been verified, see [56, 57] and [236]. Many of the critical exponents for the model have now been calculated rigorously, namely

$$(5.45) \qquad \beta = \tfrac{5}{36}, \ \gamma = \tfrac{43}{18}, \ \nu = \tfrac{4}{3}, \ \rho = \tfrac{48}{5},$$

together with the 'two-arm' exponent $\tfrac{5}{4}$, see [163, 226]. On the other hand, it has not yet been possible to extend such results to other principal percolation models such as bond or site percolation on the square lattice (some extensions have proved possible, see [66] for example).

On a related front, Smirnov [224, 225] has proved convergence of the re-scaled cluster boundaries of the critical Ising model (respectively, the associated random-cluster model) on $\mathbb{L}^2$ to SLE_3 (respectively, $SLE_{16/3}$). This will be extended in [67] to the Ising model on any so-called isoradial graph, that is, a graph embeddable in $\mathbb{R}^2$ in such a way that the vertices of any face lie on the circumference of some circle of given radius r.

The theory of SLE will soon constitute a book in its own right[8], and

---

[8]See [160, 235].

similarly for the theory of the several scaling limits that have now been proved. These general topics are beyond the scope of the current work. We restrict ourselves here to the statement and proof of Cardy's formula for critical site percolation on the triangular lattice, and we make use of the accounts to be found in [236, 237]. See also [27, 48, 205].

We consider site percolation on the triangular lattice $\mathbb{T}$, with density $p = \frac{1}{2}$ of open (or 'black') vertices. It may be proved very much as in Theorem 5.33 that $p_c = \frac{1}{2}$ for this process (see also Section 5.8), but this fact does not appear to be directly relevant to the material that follows. It is, rather, the 'self-duality' or 'self-matching' property that counts.

Let D ($\neq \mathbb{C}$) be an open simply connected domain in $\mathbb{R}^2$; for simplicity we shall assume that its boundary ∂D is a Jordan curve. Let a, b, c be distinct points of ∂D, taken in anticlockwise order around ∂D. There exists a conformal map ϕ from D to the interior of the equilateral triangle T of $\mathbb{C}$ with vertices $A = 0$, $B = 1$, $C = e^{\pi i/3}$, and such ϕ can be extended to the boundary ∂D in such a way that it becomes a homeomorphism from $D \cup \partial D$ (respectively, ∂D) to the closed triangle T (respectively, ∂T). There exists a unique such ϕ that maps a, b, c to A, B, C, respectively. With ϕ chosen accordingly, the image $X = \phi(x)$ of a fourth point $x \in \partial D$, taken for example on the arc from b to c, lies on the arc BC of T (see Figure 5.15).

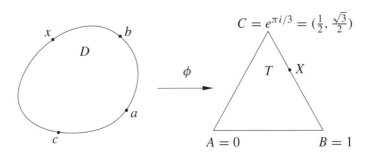

Figure 5.15. The conformal map ϕ is a bijection from D to the interior of T, and extends uniquely to the boundaries.

The triangular lattice $\mathbb{T}$ is re-scaled to have mesh-size δ, and we ask about the probability $\mathbb{P}_\delta(ac \leftrightarrow bx \text{ in } D)$ of an open path joining the arc ac to the arc bx, in an approximation to the intersection $(\delta\mathbb{T}) \cap D$ of the re-scaled lattice with D. It is a standard application of the RSW method of the last section to show that $\mathbb{P}_\delta(ac \leftrightarrow bx \text{ in } D)$ is uniformly bounded away from 0 and 1 as $\delta \to 0$. It thus seems reasonable that this probability should converge as $\delta \to 0$, and Cardy's formula (together with conformality) tells us the value of the limit.

5.46 Theorem (Cardy's formula) [59, 222, 223]. *In the notation introduced above,*

(5.47) $\mathbb{P}_\delta(ac \leftrightarrow bx$ in $D) \to |BX|$ *as* $\delta \to 0$.

Some history: In [59], Cardy stated the limit of $\mathbb{P}_\delta(ac \leftrightarrow bx$ in $D)$ in terms of a hypergeometric function of a certain cross-ratio. His derivation was based on arguments from conformal field theory. Lennart Carleson recognized the hypergeometric function in terms of the conformal map from a rectangle to a triangle, and was led to conjecture the simple form of (5.47). The limit was proved in 2001 by Stanislav Smirnov [222, 223]. The proof utilizes the three-way symmetry of the triangular lattice in a somewhat mysterious way.

Cardy's formula is, in a sense, only the beginning of the story of the scaling limit of critical two-dimensional percolation. It leads naturally to a full picture of the scaling limits of open paths, within the context of the Schramm–Löwner evolution SLE_6. While explicit application is towards the calculation of critical exponents [163, 226], SLE_6 presents a much fuller picture than this. Further details may be found in [57, 58, 223, 236]. The principal open problem at the time of writing is to extend the scaling limit beyond the site triangular model to either the bond or site model on another major lattice.

We prove Theorem 5.46 in the remainder of this section. This will be done first with $D = T$, the unit equilateral triangle, followed by the general case. Assume then that $D = T$ with T given as above. The vertices of T are $A = 0$, $B = 1$, $C = e^{\pi i/3}$. We take $\delta = 1/n$, and shall later let $n \to \infty$. Consider site percolation on $\mathbb{T}_n = (n^{-1}\mathbb{T}) \cap T$. We may draw either $\mathbb{T}_n$ or its dual graph $\mathbb{H}_n$, which comprises hexagons with centres at the vertices of $\mathbb{T}_n$, illustrated in Figure 1.5. Each vertex of $\mathbb{T}_n$ (or equivalently, each face of $\mathbb{H}_n$) is declared *black* with probability $\frac{1}{2}$, and *white* otherwise. For ease of notation later, we write $A = A_1$, $B = A_\tau$, $C = A_{\tau^2}$, where

$$\tau = e^{2\pi i/3}.$$

For vertices V, W of T we write VW for the arc of the boundary of T from V to W.

Let z be the centre of a face of $\mathbb{T}_n$ (or equivalently, $z \in V(\mathbb{H}_n)$, the vertex-set of the dual graph $\mathbb{H}_n$). The events to be studied are as follows. Let $E_1^n(z)$ be the event that there exists a self-avoiding black path from $A_1 A_\tau$ to $A_1 A_{\tau^2}$ that separates z from $A_\tau A_{\tau^2}$. Let $E_\tau^n(z)$, $E_{\tau^2}^n(z)$ be given similarly after rotating the triangle clockwise by τ and τ^2, respectively. The event $E_1^n(z)$ is illustrated in Figure 5.16. We write

$$H_j^n(z) = \mathbb{P}(E_j^n(z)), \qquad j = 1, \tau, \tau^2.$$

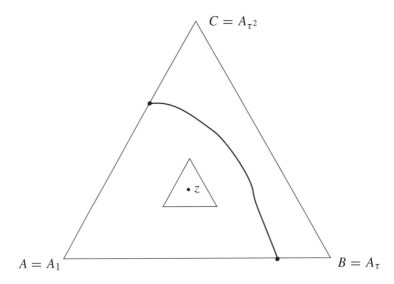

Figure 5.16. An illustration of the event $E_1^n(z)$, that z is separated from $A_\tau A_{\tau^2}$ by a black path joining $A_1 A_\tau$ and $A_1 A^{\tau^2}$.

5.48 Lemma. *The functions H_j^n, $j = 1, \tau, \tau^2$, are uniformly Hölder on $V(\mathbb{H}_n)$, in that there exist absolute constants $c \in (0, \infty)$, $\epsilon \in (0, 1)$ such that*

$$(5.49) \qquad |H_j^n(z) - H_j^n(z')| \le c|z - z'|^\epsilon, \qquad z, z' \in V(\mathbb{H}_n),$$

$$(5.50) \qquad 1 - H_j^n(z) \le c|z - A_j|^\epsilon, \qquad z \in V(\mathbb{H}_n),$$

where A_j is interpreted as the complex number at the vertex A_j.

The domain of the H_j^n may be extended as follows: the set $V(\mathbb{H}_n)$ may be viewed as the vertex-set of a triangulation of a region slightly smaller than T, on each triangle of which H_j^n may be defined by linear interpolation between its values at the three vertices. Finally, the H_j^n may be extended up to the boundary of T in such a way that the resulting functions satisfy (5.49) for all $z, z' \in T$, and

$$(5.51) \qquad H_j^n(A_j) = 1, \qquad j = 1, \tau, \tau^2.$$

Proof. It suffices to prove (5.49) for small $|z - z'|$. Suppose that $|z - z'| \le \frac{1}{100}$, say, and let F be the event that there exist both a black and a white cycle of the *entire* re-scaled triangular lattice $\mathbb{T}/n$, each of diameter smaller that $\frac{1}{4}$, and each having both z and z' in the bounded component of its complement. If F occurs, then either both or neither of the events $E_j^n(z)$, $E_j^n(z')$ occur,

whence

$$|H_j^n(z) - H_j^n(z')| \leq 1 - \mathbb{P}(F).$$

When z and z' are a 'reasonable' distance from A_j, the white cycle prevents the occurrence of one of these events without the other. The black cycle is needed when z, z' are close to A_j.

There exists $C > 0$ such that we may find $\log(C/|z - z'|)$ vertex-disjoint annuli, each containing z, z' in their central 'hole', and each within distance $\frac{1}{8}$ of both z and z' (the definition of annulus precedes Theorem 5.31). By Theorem 5.31, the chance that no such annulus contains a black (respectively, white) cycle is no greater than

$$(1 - \sigma)^{\log(C/|z-z'|)} = \left(\frac{|z - z'|}{C}\right)^{-\log(1-\sigma)},$$

whence $1 - \mathbb{P}(F) \leq c|z - z'|^\epsilon$ for suitable c and ϵ. Inequality (5.50) follows similarly with $z' = A_j$. □

It is convenient to work in the space of uniformly Hölder functions on the closed triangle T satisfying (5.49)–(5.50). By the Arzelà–Ascoli theorem (see, for example, [73, Sect. 2.4]), this space is relatively compact. Therefore, the sequence of triples $(H_1^n, H_\tau^n, H_{\tau^2}^n)$ possesses subsequential limits in the sense of uniform convergence, and we shall see that any such limit is of the form $(H_1, H_\tau, H_{\tau^2})$, where the H_j are harmonic with certain boundary conditions, and satisfy (5.49)–(5.50). The boundary conditions guarantee the uniqueness of the H_j, and it will follow that $H_j^n \to H_j$ as $n \to \infty$.

We shall see in particular that

$$H_{\tau^2}(z) = \frac{2}{\sqrt{3}}|\text{Im}(z)|,$$

the re-scaled imaginary part of z. The values of H_1 and H_τ are found by rotation. The claim of the theorem will follow by letting $z \to X \in BC$.

Let $(H_1, H_\tau, H_{\tau^2})$ be a subsequential limit as above. That the H_j are harmonic will follow from the fact that the functions

(5.52) $\qquad G_1 = H_1 + H_\tau + H_{\tau^2}, \qquad G_2 = H_1 + \tau H_\tau + \tau^2 H_{\tau^2},$

are analytic, and this analyticity will be implied by Morera's theorem on checking that the contour integrals of G_1, G_2 around triangles of a certain form are zero. The integration step amounts to summing the $H_j(z)$ over certain z and using a cancellation property that follows from the next lemma.

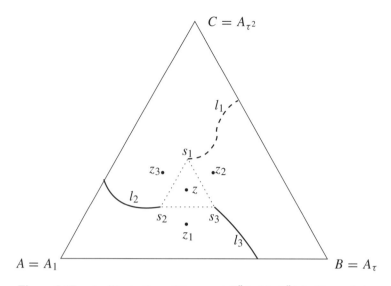

Figure 5.17. An illustration of the event $E_1^n(z_1) \setminus E_1^n(z)$. The path l_1 is white, and l_2, l_3 are black.

Let z be the centre of a face of $\mathbb{T}_n$ with vertices labelled s_1, s_2, s_3 in anticlockwise order. Let z_1, z_2, z_3 be the centres of the neighbouring faces, labelled relative to the s_j as in Figure 5.17.

5.53 Lemma. *We have that*

$$\mathbb{P}[E_1^n(z_1) \setminus E_1^n(z)] = \mathbb{P}[E_\tau^n(z_2) \setminus E_\tau^n(z)] = \mathbb{P}[E_{\tau^2}^n(z_3) \setminus E_{\tau^2}^n(z)].$$

Before proving this, we introduce the *exploration process* illustrated in Figure 5.18. Suppose that all vertices 'just outside' the arc $A_1 A_{\tau^2}$ (respectively, $A_\tau A_{\tau^2}$) of $\mathbb{T}_n$ are black (respectively, white). The exploration path is defined to be the unique path η_n on the edges of the dual (hexagonal) graph, beginning immediately above A_{τ^2} and descending to $A_1 A_\tau$ such that: as we traverse η_n from top to bottom, the vertex immediately to our left (respectively, right), looking along the path from A_{τ^2}, is white (respectively, black). When traversing η_n thus, there is a white path on the left and a black path on the right.

Proof. The event $E_1^n(z_1) \setminus E_1^n(z)$ occurs if and only if there exist disjoint paths l_1, l_2, l_3 of $\mathbb{T}_n$ such that:

(i) l_1 is white and joins s_1 to $A_\tau A_{\tau^2}$,
(ii) l_2 is black and joins s_2 to $A_1 A_{\tau^2}$,
(iii) l_3 is black and joins s_3 to $A_1 A_\tau$.

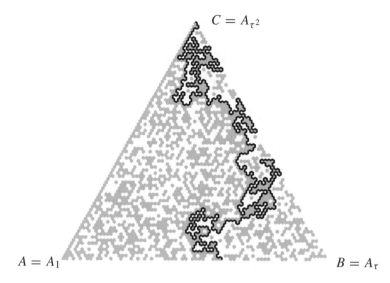

Figure 5.18. The exploration path η_n started at the top vertex A_{τ^2} and stopped when it hits the bottom side $A_1 A_\tau$ of the triangle.

See Figure 5.17 for an explanation of the notation. On this event, the exploration path η_n of Figure 5.18 passes through z and arrives at z along the edge $\langle z_3, z \rangle$ of $\mathbb{H}_n$. Furthermore, up to the time at which it hits z, it lies in the region of $\mathbb{H}_n$ between l_2 and l_1. Indeed, we may take l_2 (respectively, l_1) to be the black path (respectively, white path) of $\mathbb{T}_n$ lying on the right side (respectively, left side) of η_n up to this point.

Conditional on the event above, and with l_1 and l_2 given in terms of η_n accordingly, the states of vertices of $\mathbb{T}_n$ lying below $l_1 \cup l_2$ are independent Bernoulli variables. Thus, the conditional probability of a *black* path l_3 satisfying (iii) is the same as that of a *white* path. We make this measure-preserving change, and then we interchange the colours white/black to conclude that: $E_1^n(z_1) \setminus E_1^n(z)$ has the same probability as the event that there exist disjoint paths l_1, l_2, l_3 of $\mathbb{T}_n$ such that:

(i) l_1 is black and joins s_1 to $A_\tau A_{\tau^2}$,

(ii) l_2 is white and joins s_2 to $A_1 A_{\tau^2}$,

(iii) l_3 is black and joins s_3 to $A_1 A_\tau$.

This is precisely the event $E_\tau^n(z_2) \setminus E_\tau^n(z)$, and the lemma is proved. $\qquad\square$

We use Morera's theorem in order to show the required analyticity. This theorem states that: if $f : R \to \mathbb{C}$ is continuous on the open region R, and $\oint_\gamma f \, dz = 0$ for all closed curves γ in R, then f is analytic. It is standard (see [210, p. 208]) that it suffices to consider triangles γ in R. We may in fact restrict ourselves to equilateral triangles with one side parallel to the x-axis. This may be seen either by an approximation argument, or by an argument based on the threefold Cauchy–Riemann equations

$$(5.54) \qquad \frac{\partial f}{\partial 1} = \frac{1}{\tau} \frac{\partial f}{\partial \tau} = \frac{1}{\tau^2} \frac{\partial f}{\partial \tau^2},$$

where $\partial/\partial j$ means the derivative in the direction of the complex number j.

5.55 Lemma. *Let Γ be an equilateral triangle contained in the interior of T with sides parallel to those of T. Then*

$$\oint_{\Gamma} H_1^n(z)\, dz = \oint_{\Gamma} [H_\tau^n(z)/\tau]\, dz + O(n^{-\epsilon})$$

$$= \oint_{\Gamma} [H_{\tau^2}^n(z)/\tau^2]\, dz + O(n^{-\epsilon}),$$

where ϵ is given in Lemma 5.48.

Proof. Every triangular facet of $\mathbb{T}_n$ (that is, a triangular union of faces) points either upwards (in that its horizontal side is at its bottom) or downwards. Let Γ be an equilateral triangle contained in the interior of T with sides parallel to those of T, and assume that Γ points upwards (the same argument works for downward-pointing triangles). Let Γ^n be the subgraph of $\mathbb{T}_n$ lying within Γ, so that Γ_n is a triangular facet of $\mathbb{T}_n$. Let $\mathcal{D}^n$ be the set of downward-pointing faces of Γ^n. Let η be a vector of $\mathbb{R}^2$ such that: if z is the centre of a face of $\mathcal{D}_n$, then $z + \eta$ is the centre of a neighbouring face, that is $\eta \in \{i, i\tau, i\tau^2\}/(n\sqrt{3})$. Write

$$h_j^n(z, \eta) = \mathbb{P}[E_j^n(z + \eta) \setminus E_j^n(z)].$$

By Lemma 5.53,

$$H_1^n(z + \eta) - H_1^n(z) = h_1^n(z, \eta) - h_1^n(z + \eta, -\eta)$$

$$= h_\tau^n(z, \eta\tau) - h_\tau^n(z + \eta, -\eta\tau).$$

Now,

$$H_\tau^n(z + \eta\tau) - H_\tau^n(z) = h_\tau^n(z, \eta\tau) - h_\tau^n(z + \eta\tau, -\eta\tau),$$

and so there is a cancellation in

$$(5.56) \quad I_\eta^n = \sum_{z \in \mathcal{D}^n} [H_1^n(z + \eta) - H_1^n(z)] - \sum_{z \in \mathcal{D}^n} [H_\tau^n(z + \eta\tau) - H_\tau^n(z)]$$

of all terms except those of the form $h_\tau^n(z', -\eta\tau)$ for certain z' lying in faces of $\mathbb{T}_n$ that abut $\partial\Gamma^n$. There are $O(n)$ such z', and therefore, by Lemma 5.48,

$$(5.57) \qquad |I_\eta^n| \leq O(n^{1-\epsilon}).$$

Consider the sum

$$J^n = \frac{1}{n}(I_i^n + \tau I_{i\tau}^n + \tau^2 I_{i\tau^2}^n),$$

where I_j^n is an abbreviation for the $I_{j/n\sqrt{3}}^n$ of (5.56). The terms of the form $H_j^n(z)$ in (5.56) contribute 0 to J^n, since each is multiplied by

$$(1 + \tau + \tau^2)n^{-1} = 0.$$

The remaining terms of the form $H_j^n(z + \eta)$, $H_j^n(z + \eta\tau)$ mostly disappear also, and we are left only with terms $H_j^n(z')$ for certain z' at the centre of upwards-pointing faces of $\mathbb{T}^n$ abutting $\partial\Gamma^n$. For example, the contribution from z' if its face is at the bottom (but not the corner) of Γ^n is

$$\frac{1}{n}\left[(\tau + \tau^2)H_1^n(z') - (1 + \tau)H_\tau^n(z')\right] = -\frac{1}{n}[H_1^n(z') - H_\tau^n(z')/\tau].$$

When z' is at the right (respectively, left) edge of Γ^n, we obtain the same term multiplied by τ (respectively, τ^2). Therefore,

$$\text{(5.58)} \qquad \oint_{\Gamma^n} [H_1^n(z) - H_\tau^n(z)/\tau]\, dz = -J^n + O(n^{-\epsilon}) = O(n^{-\epsilon}),$$

by (5.57), where the first $O(n^{-\epsilon})$ term covers the fact that the z in (5.58) is a continuous rather than discrete variable. Since Γ and Γ^n differ only around their boundaries, and the H_j^n are uniformly Hölder,

$$\text{(5.59)} \qquad \oint_\Gamma [H_1^n(z) - H_\tau^n(z)/\tau]\, dz = O(n^{-\epsilon})$$

and, by a similar argument,

$$\text{(5.60)} \qquad \oint_\Gamma [H_1^n(z) - H_{\tau^2}^n(z)/\tau^2]\, dz = O(n^{-\epsilon}).$$

The lemma is proved. $\qquad\qquad\qquad\qquad\qquad\qquad\qquad\qquad\qquad\qquad\qquad$ □

As remarked after the proof of Lemma 5.48, the sequence $(H_1^n, H_\tau^n, H_{\tau^2}^n)$ possesses subsequential limits, and it suffices for convergence to show that all such limits are equal. Let $(H_1, H_\tau, H_{\tau^2})$ be such a subsequential limit. By Lemma 5.55, the contour integrals of H_1, H_τ/τ, H_{τ^2}/τ^2 around any Γ are equal. Therefore, the contour integrals of the G_i in (5.52) around any Γ equal zero. By Morera's theorem [2, 210], G_1 and G_2 are analytic on the interior of T, and furthermore they may be extended by continuity to the boundary of T. In particular, G_1 is analytic and real-valued, whence G_1 is a constant. By (5.50), $G_1(z) \to 1$ as $z \to 0$, whence

$$G_1 = H_1 + H_\tau + H_{\tau^2} \equiv 1 \quad \text{on } T.$$

Therefore, the real part of G_2 satisfies

(5.61) $\text{Re}(G_2) = H_1 - \frac{1}{2}(H_\tau + H_{\tau^2}) = \frac{1}{2}(3H_1 - 1)$,

and similarly

(5.62) $2\text{Re}(G_2/\tau) = 3H_\tau - 1, \quad 2\text{Re}(G_2/\tau^2) = 3H_{\tau^2} - 1$.

Since the H_j are the real parts of analytic functions, they are harmonic. It remains to verify the relevant boundary conditions, and we will concentrate on the function H_{τ^2}. There are two ways of doing this, of which the first specifies certain derivatives of the H_j along the boundary of T.

By continuity, $H_{\tau^2}(C) = 1$ and $H_{\tau^2} \equiv 0$ on AB. We claim that the horizontal derivative, $\partial H_{\tau^2}/\partial 1$, is 0 on $AC \cup BC$. Once this is proved, it follows that $H_{\tau^2}(z)$ is the unique harmonic function on T satisfying these boundary conditions, namely the function $2|\text{Im}(z)|/\sqrt{3}$. The remaining claim is proved as follows. Since G_2 is analytic, it satisfies the threefold Cauchy–Riemann equations (5.54). By (5.61)–(5.62),

(5.63) $\dfrac{\partial H_{\tau^2}}{\partial 1} = \dfrac{2}{3}\text{Re}\left(\dfrac{1}{\tau^2}\dfrac{\partial G_2}{\partial 1}\right) = \dfrac{2}{3}\text{Re}\left(\dfrac{1}{\tau^3}\dfrac{\partial G_2}{\partial \tau}\right) = \dfrac{\partial H_1}{\partial \tau}$.

Now, $H_1 \equiv 0$ on BC, and BC has gradient τ, whence the right side of (5.63)[9] equals 0 on BC. The same argument holds on AC with H_1 replaced by H_τ.

The alternative is slightly simpler, see [27]. For $z \in T$, $G_2(z)$ is a convex combination of 1, τ, τ^2, and thus maps T to the complex triangle T' with these three vertices. Furthermore, G_2 maps ∂T to $\partial T'$, and $G_2(A_j) = j$ for $j = 1, \tau, \tau^2$. Since G_2 is analytic on the interior of T, it is conformal, and there is a unique such conformal map with this boundary behaviour, namely that composed of a suitable dilation, rotation, and translation of T. This identifies G_2 uniquely, and the functions H_j also by (5.61)–(5.62).

This concludes the proof of Cardy's formula when the domain D is an equilateral triangle. The proof for general D is essentially the same, on noting that a conformal image of a harmonic function is harmonic. First, we approximate to the boundary of D by a cycle of the triangular lattice with mesh δ. That $G_1 (\equiv 1)$ and G_2 are analytic is proved as before, and hence the corresponding limit functions H_1, H_τ, H_{τ^2} are each harmonic with appropriate boundary conditions. We now apply conformal invariance. By the Riemann mapping theorem, there exists a conformal map ϕ from the inside of D to the inside of T that may be extended uniquely to their

---

[9]We need also that G_2 may be continued analytically beyond the boundary of T, see [237].

boundaries, and that maps a (respectively, b, c) to A (respectively, B, C). The triple $(H_1 \circ \phi^{-1}, H_\tau \circ \phi^{-1}, H_{\tau^2} \circ \phi^{-1})$ solves the corresponding problem on T. We have seen that there is a unique such triple on T, given as above, and equation (5.47) is proved.

5.8 The critical probability via the sharp-threshold theorem

We use the sharp-threshold Theorem 4.81 to prove the following.

5.64 Theorem [238]. *The critical probability of site percolation on the triangular lattice satisfies $p_c = \frac{1}{2}$. Furthermore, $\theta(\frac{1}{2}) = 0$.*

This may be proved in very much the same manner as Theorem 5.33, but we choose here to use the sharp-threshold theorem. This theorem provides a convenient 'package' for obtaining the steepness of a box-crossing probability, viewed as a function of p. Other means, more elementary and discovered earlier, may be used instead. These include: Kesten's original proof [151] for bond percolation on the square lattice, Russo's 'approximate zero–one law' [213], and, most recently, the proof of Smirnov presented in Section 5.6. Sharp-thresholds were first used in [46] in the current context, and later in [47, 100, 101]. The present proof may appear somewhat shorter than that of [46].

Proof. Let $\theta(p)$ denote the percolation probability on the triangular lattice $\mathbb{T}$. We have that $\theta(\frac{1}{2}) = 0$, just as in the proof of the corresponding lower bound for the critical probability $p_c(\mathbb{L}^2)$ in Theorem 5.33, and we say no more about this. Therefore, $p_c \geq \frac{1}{2}$.

Two steps remain. First, we shall use the sharp-threshold theorem to deduce that, when $p > \frac{1}{2}$, long rectangles are traversed with high probability in the long direction. Then we shall use that fact, within a block argument, to show that $\theta(p) > 0$.

Each vertex is declared *black* (or open) with probability p, and *white* otherwise. In the notation introduced just prior to Lemma 5.28, let $H_n = H_{16n,n\sqrt{3}}$ be the event that the rectangle $R_n = R_{16n,n\sqrt{3}}$ is traversed by a black path in the long direction. By Lemmas 5.28–5.30, there exists $\tau > 0$ such that

$$(5.65) \qquad \mathbb{P}_{\frac{1}{2}}(H_n) \geq \tau, \qquad n \geq 1.$$

Let x be a vertex of R_n, and write $I_{n,p}(x)$ for the influence of x on the event H_n under the measure $\mathbb{P}_p$, see (4.27). Now, x is pivotal for H_n if and only if:

(i) the left and right sides of R_n are connected by a black path when x is black,

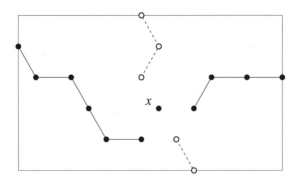

Figure 5.19. The vertex x is pivotal for H_n if and only if: there is left–right black crossing of R_n when x is black, and a top–bottom white crossing when x is white.

  (ii) the top and bottom sides of R_n are connected by a white path when x is white.

This event is illustrated in Figure 5.19.

Let $\frac{1}{2} \le p \le \frac{3}{4}$, say. By (ii),

$$(1 - p)I_{n,p}(x) \le \mathbb{P}_{1-p}(\mathrm{rad}(C_x) \ge n),$$

where

$$\mathrm{rad}(C_x) = \max\{|y - x| : x \leftrightarrow y\}$$

is the *radius* of the cluster at x. (Here, $|z|$ denotes the graph-theoretic distance from z to the origin.) Since $p \ge \frac{1}{2}$,

$$\mathbb{P}_{1-p}(\mathrm{rad}(C_x) \ge n) \le \eta_n,$$

where

$$(5.66) \qquad \eta_n = \mathbb{P}_{\frac{1}{2}}(\mathrm{rad}(C_0) \ge n) \to 0 \qquad \text{as } n \to \infty,$$

by the fact that $\theta(\frac{1}{2}) = 0$.

  By (5.65) and Theorem 4.81, for large n,

$$\mathbb{P}'_p(H_n) \ge c\tau(1 - \mathbb{P}_p(H_n)) \log[1/(8\eta_n)], \qquad p \in [\tfrac{1}{2}, \tfrac{3}{4}],$$

which may be integrated to give

$$(5.67) \qquad 1 - \mathbb{P}_p(H_n) \le (1 - \tau)[8\eta_n]^{c\tau(p-\frac{1}{2})}, \qquad p \in [\tfrac{1}{2}, \tfrac{3}{4}].$$

Let $p > \frac{1}{2}$. By (5.66)–(5.67),

$$(5.68) \qquad \mathbb{P}_p(H_n) \to 1 \qquad \text{as } n \to \infty.$$

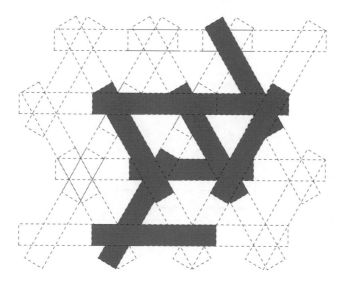

Figure 5.20. Each block is red with probability ρ_n. There is an infinite cluster of red blocks with strictly positive probability, and any such cluster contains an infinite open cluster of the original lattice.

We turn to the required block argument, which differs from that of Section 5.6 in that we shall use no explicit estimate of $\mathbb{P}_p(H_n)$. Roughly speaking, copies of the rectangle R_n are distributed about $\mathbb{T}$ in such a way that each copy corresponds to an edge of a re-scaled copy of $\mathbb{T}$. The detailed construction of this 'renormalized block lattice' is omitted from these notes, and we shall rely on Figure 5.20 for explanation. The 'blocks' (that is, the copies of R_n) are in one–one correspondence with the edges of $\mathbb{T}$, and thus we may label the blocks as B_e, $e \in \mathbb{E}_{\mathbb{T}}$. Each block intersects just ten other blocks.

Next we define a 'block event', that is, a certain event defined on the configuration within a block. The first requirement for this event is that the block be traversed by an open path in the long direction. We shall require some further paths in order that the union of two intersecting blocks contains a single component that traverses each block in its long direction. In specific, we require open paths traversing the block in the short direction, within each of the two extremal $3n \times n\sqrt{3}$ regions of the block. A block is coloured *red* if the above paths exist within it. See Figure 5.21. If two red blocks, B_e and B_f say, are such that e and f share a vertex, then their union possesses a single open component containing paths traversing each of B_e and B_f.

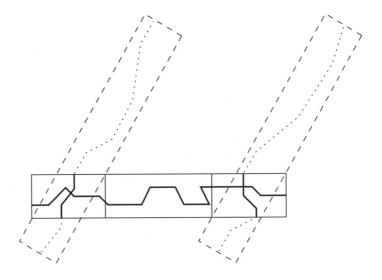

Figure 5.21. A block is declared 'red' if it contains open paths that:
(i) traverse it in the long direction, and (ii) traverse it in the short direction
within the $3n \times n\sqrt{3}$ region at each end of the block. The shorter crossings
exist if the inclined blocks are traversed in the long direction.

If the block R_n fails to be red, then one or more of the blocks in Figure
5.21 is not traversed by an open path in the long direction. Therefore,
$\rho_n := \mathbb{P}_p(R_n$ is red) satisfies

$$(5.69) \qquad 1 - \rho_n \le 3[1 - \mathbb{P}_p(H_n)] \to 0 \qquad \text{as } n \to \infty,$$

by (5.68).

The states of different blocks are dependent random variables, but any
collection of disjoint blocks have independent states. We shall count paths
in the dual, as in (3.8), to obtain that there exists, with strictly positive
probability, an infinite path in $\mathbb{T}$ comprising edges e such that every such B_e
is red. This implies the existence of an infinite open cluster in the original
lattice.

If the red cluster at the origin of the block lattice is finite, there exists a
path in the dual lattice (a copy of the hexagonal lattice) that crosses only
non-red blocks (as in Figure 3.1). Within any dual path of length m, there
exists a set of $\lfloor m/12 \rfloor$ or more edges such that the corresponding blocks
are pairwise disjoint. Therefore, the probability that the origin of the block

lattice lies in a *finite* cluster only of red blocks is no greater than

$$\sum_{m=6}^{\infty} 3^m (1 - \rho_n)^{\lfloor m/12 \rfloor}.$$

By (5.69), this may be made smaller than $\frac{1}{2}$ by choosing n sufficiently large. Therefore, $\theta(p) > 0$ for $p > \frac{1}{2}$, and the theorem is proved. $\qquad\square$

5.9 Exercises

5.1 [35] Consider bond percolation on $\mathbb{L}^2$ with $p = \frac{1}{2}$, and define the radius of the open cluster C at the origin by $\mathrm{rad}(C) = \max\{n : 0 \leftrightarrow \partial[-n,n]^2\}$. Use the BK inequality to show that

$$\mathbb{P}_{\frac{1}{2}}\left(\mathrm{rad}(C) \geq n\right) \geq \frac{1}{2\sqrt{n}}.$$

5.2 Let D_n be the largest diameter (in the sense of graph theory) of the open clusters of bond percolation on $\mathbb{Z}^d$ that intersect the box $[-n,n]^d$. Show when $p < p_c$ that $D_n/\log n \to \alpha(p)$ almost surely and in L^p, for some $\alpha(p) \in (0,\infty)$.

5.3 Consider bond percolation on $\mathbb{L}^2$ with density p. Let T_n be the box $[0,n]^2$ with periodic boundary conditions, that is, we identify any pair (u,v), (x,y) satisfying: either $u = 0, x = n, v = y$, or $v = 0, y = n, u = x$. For given $m < n$, let A be the event that there exists some translate of $[0,m]^2$ in T_n that is crossed by an open path either from top to bottom, or from left to right. Using the theory of influence or otherwise, show that

$$1 - \mathbb{P}_p(A) \leq \left[(2n^2)^{c(p-\frac{1}{2})}(2^{\lfloor n/m \rfloor^2} - 1)\right]^{-1}, \qquad p > \frac{1}{2}.$$

5.4 Consider site percolation on the triangular lattice $\mathbb{T}$, and let $\Lambda(n)$ be the ball of radius n centred at the origin. Use the RSW theorem to show that

$$\mathbb{P}_{\frac{1}{2}}(0 \leftrightarrow \partial\Lambda(n)) \geq cn^{-\alpha}, \qquad n \geq 1,$$

for constants $c, \alpha > 0$.

Using the coupling of Section 3.3 or otherwise, deduce that $\theta(p) \leq c'(p-\frac{1}{2})^\beta$ for $p > \frac{1}{2}$ and constants $c', \beta > 0$.

5.5 By adapting the arguments of Section 5.5 or otherwise, develop an RSW theory for bond percolation on $\mathbb{Z}^2$.

5.6 Let D be an open simply connected domain in $\mathbb{R}^2$ whose boundary ∂D is a Jordan curve. Let a, b, x, c be distinct points on ∂D taken in anticlockwise order. Let $\mathbb{P}_\delta(ac \leftrightarrow bx)$ be the probability that, in site percolation on the re-scaled triangular lattice $\delta\mathbb{T}$ with density $\frac{1}{2}$, there exists an open path within $D \cup \partial D$ from some point on the arc ac to some point on bx. Show that $\mathbb{P}_\delta(ac \leftrightarrow bx)$ is uniformly bounded away from 0 and 1 as $\delta \to 0$.

5.7 Let $f : D \to \mathbb{C}$, where D is an open simply connected region of the complex plane. If f is C^1 and satisfies the threefold Cauchy–Riemann equations (5.54), show that f is analytic.

6

Contact process

The contact process is a model for the spread of disease about the vertices of a graph. It has a property of duality that arises through the reversal of time. For a vertex-transitive graph such as the d-dimensional lattice, there is a multiplicity of invariant measures if and only if there is a strictly positive probability of an unbounded path of infection in space–time from a given vertex. This observation permits the use of methodology developed for the study of oriented percolation. When the underlying graph is a tree, the model has three distinct phases, termed extinction, weak survival, and strong survival. The continuous-time percolation model differs from the contact process in that the time axis is undirected.

6.1 Stochastic epidemics

One of the simplest stochastic models for the spread of an epidemic is as follows. Consider a population of constant size $N + 1$ that is suffering from an infectious disease. We can model the spread of the disease as a Markov process. Let $X(t)$ be the number of healthy individuals at time t and suppose that $X(0) = N$. We assume that, if $X(t) = n$, the probability of a new infection during a short time-interval $(t, t + h)$ is proportional to the number of possible encounters between ill folk and healthy folk. That is,

$$\mathbb{P}\big(X(t + h) = n - 1 \,\big|\, X(t) = n\big) = \lambda n(N + 1 - n)h + o(h) \quad \text{as } h \downarrow 0.$$

In the simplest situation, we assume that nobody recovers. It is easy to show that

$$G(s, t) = \mathbb{E}(s^{X(t)}) = \sum_{n=0}^{N} s^n \mathbb{P}(X(t) = n)$$

satisfies

$$\frac{\partial G}{\partial t} = \lambda(1 - s)\left(N\frac{\partial G}{\partial s} - s\frac{\partial^2 G}{\partial s^2}\right)$$

with $G(s, 0) = s^N$. There is no simple way of solving this equation, although a lot of information is available about approximate solutions.

This epidemic model is over-simplistic through the assumptions that:
- the process is Markovian,
- there are only two states and no recovery,
- there is total mixing, in that the rate of spread is proportional to the product of the numbers of infectives and susceptibles.

In 'practice' (computer viruses apart), an individual infects only others in its immediate (bounded) vicinity. The introduction of spatial relationships into such a model adds a major complication, and leads to the so-called 'contact model' of Harris [136]. In the contact model, the members of the population inhabit the vertex-set of a given graph. Infection takes place between neighbours, and recovery is permitted.

Let $G = (V, E)$ be a (finite or infinite) graph with bounded vertex-degrees. The contact model on G is a continuous-time Markov process on the state space $\Sigma = \{0, 1\}^V$. A state is therefore a 0/1 vector $\xi = (\xi(x) : x \in V)$, where 0 represents the healthy state and 1 the ill state. There are two parameters: an infection rate λ and a recovery rate δ. Transition-rates are given informally as follows. Suppose that the state at time t is $\xi \in \Sigma$, and let $x \in V$. Then

$$\mathbb{P}(\xi_{t+h}(x) = 0 \mid \xi_t = \xi) = \delta h + \mathrm{o}(h), \qquad \text{if } \xi(x) = 1,$$
$$\mathbb{P}(\xi_{t+h}(x) = 1 \mid \xi_t = \xi) = \lambda N_\xi(x) h + \mathrm{o}(h), \qquad \text{if } \xi(x) = 0,$$

where $N_\xi(x)$ is the number of neighbours of x that are infected in ξ,

$$N_\xi(x) = \left|\{y \in V : y \sim x, \, \xi(y) = 1\}\right|.$$

Thus, each ill vertex recovers at rate δ, and in the meantime infects any given neighbour at rate λ.

Care is needed when specifying a Markov process through its transition rates, especially when G is infinite, since then Σ is uncountable. We shall see in the next section that the contact model can be constructed via a countably infinite collection of Poisson processes. More general approaches to the construction of interacting particle processes are described in [167] and summarized in Section 10.1.

6.2 Coupling and duality

The contact model can be constructed in terms of families of Poisson processes. This representation is both informative and useful for what follows. For each $x \in V$, we draw a 'time-line' $[0, \infty)$. On the time-line $\{x\} \times [0, \infty)$

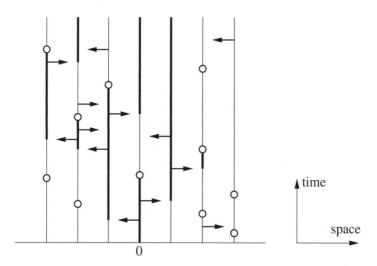

Figure 6.1. The so-called 'graphical representation' of the contact process on the line $\mathbb{L}$. The horizontal line represents 'space', and the vertical line above a point x is the time-line at x. The marks ∘ are the points of cure, and the arrows are the arrows of infection. Suppose we are told that, at time 0, the origin is the unique infected point. In this picture, the initial infective is marked 0, and the bold lines indicate the portions of space–time that are infected.

we place a Poisson point process D_x with intensity δ. For each *ordered* pair $x, y \in V$ of neighbours, we let $B_{x,y}$ be a Poisson point process with intensity λ. These processes are taken to be independent of each other, and we assume without loss of generality that the times occurring in the processes are distinct. Points in each D_x are called 'points of cure', and points in $B_{x,y}$ are called 'arrows of infection' from x to y. The appropriate probability measure is denoted by $\mathbb{P}_{\lambda,\delta}$.

The situation is illustrated in Figure 6.1 with $G = \mathbb{L}$. Let $(x, s), (y, t) \in V \times [0, \infty)$ where $s \leq t$. We define a *(directed) path* from (x, s) to (y, t) to be a sequence $(x, s) = (x_0, t_0), (x_0, t_1), (x_1, t_1), (x_1, t_2), \ldots, (x_n, t_{n+1}) = (y, t)$ with $t_0 \leq t_1 \leq \cdots \leq t_{n+1}$, such that:

(i) each interval $\{x_i\} \times [t_i, t_{i+1}]$ contains no points of D_{x_i},

(ii) $t_i \in B_{x_{i-1}, x_i}$ for $i = 1, 2, \ldots, n$.

We write $(x, s) \to (y, t)$ if there exists such a directed path.

We think of a point (x, u) of cure as meaning that an infection at x just prior to time u is cured at time u. An arrow of infection from x to y at time u means that an infection at x just prior to u is passed at time u to y. Thus,

$(x, s) \rightarrow (y, t)$ means that y is infected at time t if x is infected at time s.

Let $\xi_0 \in \Sigma = \{0, 1\}^V$, and define $\xi_t \in \Sigma$, $t \in [0, \infty)$, by $\xi_t(y) = 1$ if and only if there exists $x \in V$ such that $\xi_0(x) = 1$ and $(x, 0) \rightarrow (y, t)$. It is clear that $(\xi_t : t \in [0, \infty))$ is a contact model with parameters λ and δ.

The above 'graphical representation' has several uses. First, it is a geometrical picture of the spread of infection providing a coupling of contact models with all possible initial configurations ξ_0. Secondly, it provides couplings of contact models with different λ and δ, as follows. Let $\lambda_1 \leq \lambda_2$ and $\delta_1 \geq \delta_2$, and consider the above representation with $(\lambda, \delta) = (\lambda_2, \delta_1)$. If we remove each point of cure with probability δ_2/δ_1 (respectively, each arrow of infection with probability λ_1/λ_2), we obtain a representation of a contact model with parameters (λ_2, δ_2) (respectively, parameters (λ_1, δ_1)). We obtain thus that the passage of infection is non-increasing in δ and non-decreasing in λ.

There is a natural one–one correspondence between Σ and the power set 2^V of the vertex-set, given by $\xi \leftrightarrow I_\xi = \{x \in V : \xi(x) = 1\}$. We shall frequently regard vectors ξ as sets I_ξ. For $\xi \in \Sigma$ and $A \subseteq V$, we write ξ_t^A for the value of the contact model at time t starting at time 0 from the set A of infectives. It is immediate by the rules of the above coupling that:

(a) the coupling is *monotone* in that $\xi_t^A \subseteq \xi_t^B$ if $A \subseteq B$,

(b) moreover, the coupling is *additive* in that $\xi_t^{A \cup B} = \xi_t^A \cup \xi_t^B$.

6.1 Theorem (Duality relation). *For A, $B \subseteq V$,*

(6.2) $\mathbb{P}_{\lambda,\delta}(\xi_t^A \cap B \neq \varnothing) = \mathbb{P}_{\lambda,\delta}(\xi_t^B \cap A \neq \varnothing)$.

Equation (6.2) can be written in the form

$$\mathbb{P}_{\lambda,\delta}^A(\xi_t \equiv 0 \text{ on } B) = \mathbb{P}_{\lambda,\delta}^B(\xi_t \equiv 0 \text{ on } A),$$

where the superscripts indicate the initial states. This may be termed 'weak' duality, in that it involves probabilities. There is also a 'strong' duality involving configurations of the graphical representation, that may be expressed informally as follows. Suppose we reverse the direction of time in the 'primary' graphical representation, and also the directions of the arrows. The law of the resulting process is the same as that of the original. Furthermore, any path of infection in the primary process, after reversal, becomes a path of infection in the reversed process.

Proof. The event on the left side of (6.2) is the union over $a \in A$ and $b \in B$ of the event that $(a, 0) \rightarrow (b, t)$. If we reverse the direction of time and the directions of the arrows of infection, the probability of this event is unchanged and it corresponds now to the event on the right side of (6.2). $\square$

6.3 Invariant measures and percolation

In this and the next section, we consider the contact process $\xi = (\xi_t : t \geq 0)$ on the d-dimensional cubic lattice $\mathbb{L}^d$ with $d \geq 1$. Thus, ξ is a Markov process on the state space $\Sigma = \{0, 1\}^{\mathbb{Z}^d}$. Let $\mathcal{I}$ be the set of invariant measures of ξ, that is, the set of probability measures μ on Σ such that $\mu S_t = \mu$, where $S = (S_t : t \geq 0)$ is the transition semigroup of the process.[1] It is elementary that $\mathcal{I}$ is a convex set of measures: if $\phi_1, \phi_2 \in \mathcal{I}$, then $\alpha\phi_1 + (1 - \alpha)\phi_2 \in \mathcal{I}$ for $\alpha \in [0, 1]$. Therefore, $\mathcal{I}$ is determined by knowledge of the set $\mathcal{I}_e$ of *extremal* invariant measures.

The partial order on Σ induces a partial order on probability measures on Σ in the usual way, and we denote this by $\leq_{\text{st}}$. It turns out that $\mathcal{I}$ possesses a 'minimal' and 'maximal' element, with respect to $\leq_{\text{st}}$. The minimal measure (or 'lower invariant measure') is the measure, denoted $\delta_\varnothing$, that places probability 1 on the empty set. It is called 'lower' because $\delta_\varnothing \leq_{\text{st}} \mu$ for all measures μ on Σ.

The maximal measure (or 'upper invariant measure') is constructed as the weak limit of the contact model beginning with the set $\xi_0 = \mathbb{Z}^d$. Let μ_s denote the law of $\xi_s^{\mathbb{Z}^d}$. Since $\xi_s^{\mathbb{Z}^d} \subseteq \mathbb{Z}^d$,

$$\mu_0 S_s = \mu_s \leq_{\text{st}} \mu_0.$$

By the monotonicity of the coupling,

$$\mu_{s+t} = \mu_0 S_s S_t = \mu_s S_t \leq_{\text{st}} \mu_0 S_t = \mu_t,$$

whence the limit

$$\lim_{t \to \infty} \mu_t(f)$$

exists for any bounded increasing function $f : \Sigma \to \mathbb{R}$. It is a general result of measure theory that the space of probability measures on a compact sample space is relatively compact (see [39, Sect. 1.6] and [73, Sect. 9.3]). The space $(\Sigma, \mathcal{F})$ is indeed compact, whence the weak limit

$$\overline{\nu} = \lim_{t \to \infty} \mu_t$$

exists. Since $\overline{\nu}$ is a limiting measure for the Markov process, it is invariant, and it is called the *upper invariant measure*. It is clear by the method of its construction that $\overline{\nu}$ is invariant under the action of any translation of $\mathbb{L}^d$.

6.3 Proposition. *We have that* $\delta_\varnothing \leq_{\text{st}} \nu \leq_{\text{st}} \overline{\nu}$ *for every* $\nu \in \mathcal{I}$.

---

[1] A discussion of the transition semigroup and its relationship to invariant measures can be found in Section 10.1. The semigroup S is Feller, see the footnote on page 191.

Proof. Let $v \in \mathit{I}$. The first inequality is trivial. Clearly, $v \leq_{\mathrm{st}} \mu_0$, since μ_0 is concentrated on the maximal set $\mathbb{Z}^d$. By the monotonicity of the coupling,

$$v = vS_t \leq_{\mathrm{st}} \mu_0 S_t = \mu_t, \qquad t \geq 0.$$

Let $t \to \infty$ to obtain that $v \leq_{\mathrm{st}} \overline{v}$. $\square$

By Proposition 6.3, there exists a unique invariant measure if and only if $\overline{v} = \delta_{\varnothing}$. In order to understand when this is so, we deviate briefly to consider a percolation-type question. Suppose we begin the process at a singleton, the origin say, and ask whether the probability of survival for all time is strictly positive. That is, we work with the percolation-type probability

(6.4) $\theta(\lambda, \delta) = \mathbb{P}_{\lambda,\delta}(\xi_t^0 \neq \varnothing \text{ for all } t \geq 0),$

where $\xi_t^0 = \xi_t^{\{0\}}$. By a re-scaling of time, $\theta(\lambda, \delta) = \theta(\lambda/\delta, 1)$, and *we assume henceforth in his section that $\delta = 1$, and we write $\mathbb{P}_\lambda = \mathbb{P}_{\lambda,1}$.*

6.5 Proposition. *The density of ill vertices under $\overline{v}$ equals $\theta(\lambda)$. That is,*

$$\theta(\lambda) = \overline{v}(\{\sigma \in \Sigma : \sigma_x = 1\}), \qquad x \in \mathbb{Z}^d.$$

Proof. The event $\{\xi_T^0 \cap \mathbb{Z}^d \neq \varnothing\}$ is non-increasing in T, whence

$$\theta(\lambda) = \lim_{T \to \infty} \mathbb{P}_\lambda(\xi_T^0 \cap \mathbb{Z}^d \neq \varnothing).$$

By Proposition 6.1,

$$\mathbb{P}_\lambda(\xi_T^0 \cap \mathbb{Z}^d \neq \varnothing) = \mathbb{P}_\lambda(\xi_T^{\mathbb{Z}^d}(0) = 1),$$

and by weak convergence,

$$\mathbb{P}_\lambda(\xi_T^{\mathbb{Z}^d}(0) = 1) \to \overline{v}(\{\sigma \in \Sigma : \sigma_0 = 1\}).$$

The claim follows by the translation-invariance of $\overline{v}$. $\square$

We define the critical value of the process by

$$\lambda_c = \lambda_c(d) = \sup\{\lambda : \theta(\lambda) = 0\}.$$

The function $\theta(\lambda)$ is non-decreasing, so that

$$\theta(\lambda) \begin{cases} = 0 & \text{if } \lambda < \lambda_c, \\ > 0 & \text{if } \lambda > \lambda_c. \end{cases}$$

By Proposition 6.5,

$$\overline{v} \begin{cases} = \delta_\varnothing & \text{if } \lambda < \lambda_c, \\ \neq \delta_\varnothing & \text{if } \lambda > \lambda_c. \end{cases}$$

The case $\lambda = \lambda_c$ is delicate, especially when $d \geq 2$, and it has been shown in [36], using a slab argument related to that of the proof of Theorem 5.17, that $\theta(\lambda_c) = 0$ for $d \geq 1$. We arrive at the following characterization of uniqueness of extremal invariant measures.

6.6 Theorem [36]. *Consider the contact model on $\mathbb{L}^d$ with $d \geq 1$. The set $\mathcal{I}$ of invariant measures comprises a singleton if and only if $\lambda \leq \lambda_c$. That is, $\mathcal{I} = \{\delta_\varnothing\}$ if and only if $\lambda \leq \lambda_c$.*

There are further consequences of the arguments of [36] of which we mention one. The geometrical constructions of [36] enable a proof of the equivalent for the contact model of the 'slab' percolation Theorem 5.17. This in turn completes the proof, initiated in [76, 80], that the set of extremal invariant measures of the contact model on $\mathbb{L}^d$ is exactly $\mathcal{I}_e = \{\delta_\varnothing, \overline{\nu}\}$. See [78] also.

6.4 The critical value

This section is devoted to the following theorem.[2] Recall that the rate of cure is taken as $\delta = 1$.

6.7 Theorem [136]. *For $d \geq 1$, we have that $(2d)^{-1} < \lambda_c(d) < \infty$.*

The lower bound is easily improved to $\lambda_c(d) \geq (2d - 1)^{-1}$. The upper bound may be refined to $\lambda_c(d) \leq d^{-1}\lambda_c(1) < \infty$, as indicated in Exercise 6.2. See the accounts of the contact model in the two volumes [167, 169] by Tom Liggett.

Proof. The lower bound is obtained by a random walk argument that is sketched here.[3] The integer-valued process $N_t = |\xi_t^0|$ decreases by 1 at rate N_t. It increases by 1 at rate λT_t, where T_t is the number of edges of $\mathbb{L}^d$ exactly one of whose endvertices x satisfies $\xi_t^0(x) = 1$. Now, $T_t \leq 2d N_t$, and so the jump-chain of N_t is bounded above by a simple random walk $R = (R_n : n \geq 0)$ on $\{0, 1, 2, \ldots\}$, with absorption at 0, and that moves to the right with probability

$$p = \frac{2d\lambda}{1 + 2d\lambda}$$

at each step. It is elementary that

$$\mathbb{P}(R_n = 0 \text{ for some } n \geq 0) = 1 \quad \text{if} \quad p \leq \tfrac{1}{2},$$

and it follows that

$$\theta(\lambda) = 0 \quad \text{if} \quad \lambda < \frac{1}{2d}.$$

---

[2]There are physical reasons to suppose that $\lambda_c(1) = 1.6494\ldots$, see the discussion of the so-called reggeon spin model in [102, 167].

[3]The details are left as an exercise.

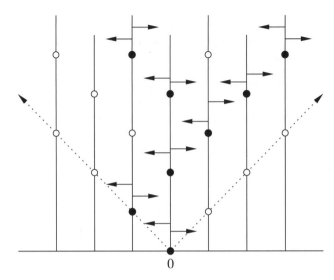

Figure 6.2. The points $(m, n\Delta)$ are marked for even values of $m + n$. A point is 'open' if there are arrows of infection immediately following, and no point of cure just prior. The open points form a site percolation process on the rotated positive quadrant of the square lattice.

Just as in the case of percolation (Theorem 3.2) the upper bound on λ_c requires more work. Since $\mathbb{L}$ may be viewed as a subgraph of $\mathbb{L}^d$, it is elementary that $\lambda_c(d) \le \lambda_c(1)$. We show by a discretization argument that $\lambda_c(1) < \infty$. Let $\Delta > 0$, and let $m, n \in \mathbb{Z}$ be such that $m + n$ is even. We shall define independent random variables $X_{m,n}$ taking the values 0 and 1. We declare $X_{m,n} = 1$, and call (m, n) *open*, if and only if, in the graphical representation of the contact model, the following two events occur:

(a) there is no point of cure in the interval $\{m\} \times \big((n-1)\Delta, (n+1)\Delta\big]$,

(b) there exist left and right pointing arrows of infection from the interval $\{m\} \times \big(n\Delta, (n+1)\Delta\big]$.

(See Figure 6.2.) It is immediate that the $X_{m,n}$ are independent, and

$$p = p(\Delta) = \mathbb{P}_\lambda(X_{m,n} = 1) = e^{-2\Delta}(1 - e^{-\lambda\Delta})^2.$$

We choose Δ to maximize $p(\Delta)$, which is to say that

$$e^{-\lambda\Delta} = \frac{1}{1+\lambda},$$

and

(6.8) $$p = \frac{\lambda^2}{(1+\lambda)^{2+2/\lambda}}.$$

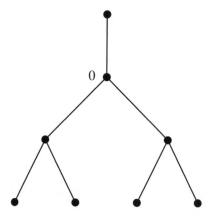

Figure 6.3. Part of the binary tree T_2.

Consider the $X_{m,n}$ as giving rise to a directed site percolation model on the first quadrant of a rotated copy of $\mathbb{L}^2$. It can be seen that $\xi^0_{n\Delta} \supseteq B_n$, where B_n is the set of vertices of the form (m, n) that are reached from $(0, 0)$ along open paths of the percolation process. Now,

$$\mathbb{P}_\lambda\big(|B_n| = \infty \text{ for all } n \geq 0\big) > 0 \quad \text{if} \quad p > \vec{p}^{\,\text{site}}_{\text{c}},$$

where $\vec{p}^{\,\text{site}}_{\text{c}}$ is the critical probability of the percolation model. By (6.8),

$$\theta(\lambda) > 0 \quad \text{if} \quad \frac{\lambda^2}{(1+\lambda)^{2+2/\lambda}} > \vec{p}^{\,\text{site}}_{\text{c}}.$$

Since[4] $\vec{p}^{\,\text{site}}_{\text{c}} < 1$, the final inequality is valid for sufficiently large λ, and we deduce that $\lambda_{\text{c}}(1) < \infty$. $\qquad\qquad\qquad\qquad\qquad\qquad\qquad\qquad\square$

6.5 The contact model on a tree

Let $d \geq 2$ and let T_d be the homogeneous (infinite) labelled tree in which every vertex has degree $d + 1$, illustrated in Figure 6.3. We identify a distinguished vertex, called the *origin* and denoted 0. Let $\xi = (\xi_t : t \geq 0)$ be a contact model on T_d with infection rate λ and initial state $\xi_0 = \{0\}$, and take $\delta = 1$.

With

$$\theta(\lambda) = \mathbb{P}_\lambda(\xi_t \neq \varnothing \text{ for all } t),$$

[4]*Exercise.*

the process is said to *die out* if $\theta(\lambda) = 0$, and to *survive* if $\theta(\lambda) > 0$. It is said to *survive strongly* if

$$\mathbb{P}_\lambda\big(\xi_t(0) = 1 \text{ for unbounded times } t\big) > 0,$$

and to *survive weakly* if it survives but it does not survive strongly. A process that survives weakly has the property that (with strictly positive probability) the illness exists for all time, but that (almost surely) there is a final time at which any given vertex is infected. It can be shown that weak survival never occurs on a lattice $\mathbb{L}^d$, see [169]. The picture is quite different on a tree.

The properties of survival and strong survival are evidently non-decreasing in λ, whence there exist values λ_c, λ_{ss} satisfying $\lambda_c \leq \lambda_{ss}$ such that the process

$$\begin{aligned}
\text{dies out} &\quad \text{if} \quad \lambda < \lambda_c, \\
\text{survives weakly} &\quad \text{if} \quad \lambda_c < \lambda < \lambda_{ss}, \\
\text{survives strongly} &\quad \text{if} \quad \lambda > \lambda_{ss}.
\end{aligned}$$

When is it the case that $\lambda_c < \lambda_{ss}$? The next theorem indicates that this occurs on T_d if $d \geq 6$. It was further proved in [196] that strict inequality holds whenever $d \geq 3$, and this was extended in [168] to $d \geq 2$. See [169, Chap. I.4] and the references therein.

6.9 Theorem [196]. *For the contact model on the tree T_d with $d \geq 2$,*

$$\frac{1}{2\sqrt{d}} \leq \lambda_c < \frac{1}{d-1}.$$

Proof. First we prove the upper bound. Let $\rho \in (0, 1)$, and $\nu_\rho(A) = \rho^{|A|}$ for any finite subset A of the vertex-set V of T_d. We shall observe the process $\nu_\rho(\xi_t)$. Let $g^A(t) = \mathbb{E}_\lambda^A(\nu_\rho(\xi_t))$. It is an easy calculation that

(6.10) $$g^A(t) = |A|t\left[\frac{\nu_\rho(A)}{\rho}\right] + \lambda N_A t\big[\rho\nu_\rho(A)\big]$$
$$+ (1 - |A|t - \lambda N_A t)\nu_\rho(A) + \mathrm{o}(t),$$

as $t \downarrow 0$, where

$$N_A = \big|\{\langle x, y\rangle : x \in A, \; y \notin A\}\big|$$

is the number of edges of T_d with exactly one endvertex in A. Now,

(6.11) $$N_A \geq (d+1)|A| - 2(|A| - 1),$$

since there are no more than $|A| - 1$ edges having both endvertices in A.
By (6.10),

(6.12) $$\left.\frac{d}{dt}g^A(t)\right|_{t=0} = (1-\rho)\left(\frac{|A|}{\rho} - \lambda N_A\right)v_\rho(A)$$

$$\le (1-\rho)v_\rho(A)\left[\frac{|A|}{\rho}(1-\lambda\rho(d-1)) - 2\lambda\right]$$

$$\le -2\lambda(1-\rho)v_\rho(A) \le 0,$$

whenever

(6.13) $$\lambda\rho(d-1) \ge 1.$$

Assume that (6.13) holds. By (6.12) and the Markov property,

(6.14) $$\frac{d}{du}g^A(u) = \mathbb{E}_\lambda^A\left(\left.\frac{d}{dt}g^{\xi_u}(t)\right|_{t=0}\right) \le 0,$$

implying that $g^A(u)$ is non-increasing in u.

With $A = \{0\}$, we have that $g(0) = \rho < 1$, and therefore

$$\lim_{t\to\infty}g(t) \le \rho.$$

On the other hand, if the process dies out, then (almost surely) $\xi_t = \varnothing$ for all
large t, so that, by the bounded convergence theorm, $g(t) \to 1$ as $t \to \infty$.
From this contradiction, we deduce that the process survives whenever there
exists $\rho \in (0, 1)$ such that (6.13) holds. Therefore, $(d-1)\lambda_c < 1$.

Turning to the lower bound, let $\rho \in (0, 1)$ once again. We draw the tree
in the manner of Figure 6.4, and we let $l(x)$ be the generation number of the
vertex x relative to 0 in this representation. For a finite subset A of V, let

$$w_\rho(A) = \sum_{x\in A}\rho^{l(x)},$$

with the convention that an empty summation equals 0.

As in (6.12), $h^A(t) = \mathbb{E}_\lambda^A(w_\rho(\xi_t))$ satisfies

(6.15) $$\left.\frac{d}{dt}h^A(t)\right|_{t=0} = \sum_{x\in A}\left(-\rho^{l(x)} + \lambda\sum_{\substack{y\in V:\ y\sim x,\\ y\notin A}}\rho^{l(y)}\right)$$

$$\le -w_\rho(A) + \lambda\sum_{x\in A}\rho^{l(x)}[d\rho + \rho^{-1}]$$

$$= (\lambda d\rho + \lambda\rho^{-1} - 1)w_\rho(A).$$

Set

(6.16) $$\rho = \frac{1}{\sqrt{d}}, \quad \lambda = \frac{1}{2\sqrt{d}},$$

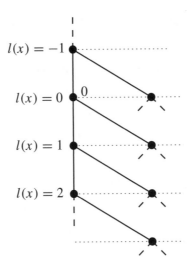

$l(x) = -1$

$l(x) = 0$

$l(x) = 1$

$l(x) = 2$

Figure 6.4. The binary tree T_2 'suspended' from a given doubly infinite path, with the generation numbers as marked.

so that $\lambda d\rho + \lambda\rho^{-1} - 1 = 0$. By (6.15), $w_\rho(\xi_t)$ is a positive supermartingale. By the martingale convergence theorem, the limit

$$(6.17) \qquad\qquad M = \lim_{t\to\infty} w_\rho(\xi_t),$$

exists $\mathbb{P}_\lambda^A$-almost surely. See [121, Sect. 12.3] for an account of the convergence of martingales.

On the event $I = \{\xi_t(0) = 1$ for unbounded times $t\}$, the process $w_\rho(\xi_t)$ changes its value (almost surely) by $\rho^0 = 1$ on an unbounded set of times t, in contradiction of (6.17). Therefore, $\mathbb{P}_\lambda^A(I) = 0$, and the process does not converge strongly under (6.16). The theorem is proved. □

6.6 Space–time percolation

The percolation models of Chapters 2 and 5 are discrete in that they inhabit a discrete graph $G = (V, E)$. There are a variety of *continuum* models of interest (see [110] for a summary) of which we distinguish the continuum model on $V \times \mathbb{R}$. We can consider this as the contact model with *undirected* time. We will encounter the related continuum random-cluster model in Chapter 9, together with its application to the quantum Ising model.

Let $G = (V, E)$ be a finite graph. The percolation model of this section inhabits the space $V \times \mathbb{R}$, which we refer to as space–time, and we consider $V \times \mathbb{R}$ as obtained by attaching a 'time-line' $(-\infty, \infty)$ to each vertex $x \in V$.

Let $\lambda, \delta \in (0, \infty)$. The continuum percolation model on $V \times \mathbb{R}$ is constructed via processes of 'cuts' and 'bridges' as follows. For each $x \in V$, we select a Poisson process D_x of points in $\{x\} \times \mathbb{R}$ with intensity δ; the processes $\{D_x : x \in V\}$ are independent, and the points in the D_x are termed 'cuts'. For each $e = \langle x, y \rangle \in E$, we select a Poisson process B_e of points in $\{e\} \times \mathbb{R}$ with intensity λ; the processes $\{B_e : e \in E\}$ are independent of each other and of the D_x. Let $\mathbb{P}_{\lambda, \delta}$ denote the probability measure associated with the family of such Poisson processes indexed by $V \cup E$.

For each $e = \langle x, y \rangle \in E$ and $(e, t) \in B_e$, we think of (e, t) as an edge joining the endpoints (x, t) and (y, t), and we refer to this edge as a 'bridge'. For $(x, s), (y, t) \in V \times \mathbb{R}$, we write $(x, s) \leftrightarrow (y, t)$ if there exists a path π with endpoints $(x, s), (y, t)$ such that: π is a union of cut-free sub-intervals of $V \times \mathbb{R}$ and bridges. For $\Lambda, \Delta \subseteq V \times \mathbb{R}$, we write $\Lambda \leftrightarrow \Delta$ if there exist $a \in \Lambda$ and $b \in \Delta$ such that $a \leftrightarrow b$.

For $(x, s) \in V \times \mathbb{R}$, let $C_{x,s}$ be the set of all points (y, t) such that $(x, s) \leftrightarrow (y, t)$. The clusters $C_{x,s}$ have been studied in [37], where the case $G = \mathbb{Z}^d$ was considered in some detail. Let 0 denote the origin $(0, 0) \in \mathbb{Z}^d \times \mathbb{R}$, and let $C = C_0$ denote the cluster at the origin. Noting that C is a union of line-segments, we write $|C|$ for its Lebesgue measure. The *radius* $\mathrm{rad}(C)$ of C is given by

$$\mathrm{rad}(C) = \sup\{\|x\| + |t| : (x, t) \in C\},$$

where

$$\|x\| = \sup_i |x_i|, \qquad x = (x_1, x_2, \ldots, x_d) \in \mathbb{Z}^d,$$

is the supremum norm on $\mathbb{Z}^d$.

The critical point of the process is defined by

$$\lambda_c(\delta) = \sup\{\lambda : \theta(\lambda, \delta) = 0\},$$

where

$$\theta(\lambda, \delta) = \mathbb{P}_{\lambda, \delta}(|C| = \infty).$$

It is immediate by re-scaling time that $\theta(\lambda, \delta) = \theta(\lambda/\delta, 1)$, and we shall use the abbreviations $\lambda_c = \lambda_c(1)$ and $\theta(\lambda) = \theta(\lambda, 1)$.

6.18 Theorem [37]. *Let $G = \mathbb{L}^d$ where $d \geq 1$, and consider continuum percolation on $\mathbb{L}^d \times \mathbb{R}$.*

(a) *Let $\lambda, \delta \in (0, \infty)$. There exist γ, ν satisfying $\gamma, \nu > 0$ for $\lambda/\delta < \lambda_c$ such that*

$$\mathbb{P}_{\lambda, \delta}(|C| \geq k) \leq e^{-\gamma k}, \qquad k > 0,$$

$$\mathbb{P}_{\lambda, \delta}(\mathrm{rad}(C) \geq k) \leq e^{-\nu k}, \qquad k > 0.$$

(b) *When $d = 1$, $\lambda_c = 1$ and $\theta(1) = 0$.*

There is a natural duality in $1+1$ dimensions (that is, when the underlying graph is the line $\mathbb{L}$), and it is easily seen in this case that the process is self-dual when $\lambda = \delta$. Part (b) identifies this self-dual point as the critical point. For general $d \geq 1$, the continuum percolation model on $\mathbb{L}^d \times \mathbb{R}$ has exponential decay of connectivity when $\lambda/\delta < \lambda_c$. The proof, which is omitted, uses an adaptation to the continuum of the methods used for $\mathbb{L}^{d+1}$. Theorem 6.18 will be useful for the study of the quantum Ising model in Section 9.4.

There has been considerable interest in the behaviour of the continuum percolation model on a graph G when the environment is itself chosen at random, that is, we take the $\lambda = \lambda_e$, $\delta = \delta_x$ to be random variables. More precisely, suppose that the Poisson process of cuts at a vertex $x \in V$ has some intensity δ_x, and that of bridges parallel to the edge $e = \langle x, y \rangle \in E$ has some intensity λ_e. Suppose further that the $\delta_x, x \in V$, are independent, identically distributed random variables, and the $\lambda_e, e \in E$ also. Write Δ and Λ for independent random variables having the respective distributions, and P for the probability measure governing the environment. [As before, $\mathbb{P}_{\lambda,\delta}$ denotes the measure associated with the percolation model in the given environment. The above use of the letters Δ, Λ to denote random variables is temporary only.] The problem of understanding the behaviour of the system is now much harder, because of the fluctuations in intensities about G.

If there exist $\lambda', \delta' \in (0, \infty)$ such that $\lambda'/\delta' < \lambda_c$ and

$$P(\Lambda \leq \lambda') = P(\Delta \geq \delta') = 1,$$

then the process is almost surely dominated by the subcritical percolation process with parameters λ', δ', whence there is (almost surely) exponential decay in the sense of Theorem 6.18(i). This can fail in an interesting way if there is no such almost-sure domination, in that (under certain conditions) we can prove exponential decay in the space-direction but only a weaker decay in the time-direction. The problem arises since there will generally be regions of space that are favourable to the existence of large clusters, and other regions that are unfavourable. In a favourable region, there may be unnaturally long connections between two points with similar values for their time-coordinates.

For $(x, s), (y, t) \in \mathbb{Z}^d \times \mathbb{R}$ and $q \geq 1$, we define

$$\delta_q(x, s; y, t) = \max\{\|x - y\|, [\log(1 + |s - t|)]^q\}.$$

6.19 Theorem [154, 155]. *Let $G = \mathbb{L}^d$, where $d \geq 1$. Suppose that*

$$K = \max \left\{ P\big([\log(1 + \Lambda)]^\beta\big), P\big([\log(1 + \Delta^{-1})]^\beta\big) \right\} < \infty,$$

for some $\beta > 2d^2\big(1 + \sqrt{1 + d^{-1}} + (2d)^{-1}\big)$. There exists $Q = Q(d, \beta) > 1$ such that the following holds. For $q \in [1, Q)$ and $m > 0$, there exists $\epsilon = \epsilon(d, \beta, K, m, q) > 0$ and $\eta = \eta(d, \beta, q) > 0$ such that: if

$$P\big([\log(1 + (\Lambda/\Delta))]^\beta\big) < \epsilon,$$

there exist identically distributed random variables $D_x \in L^\eta(P)$, $x \in \mathbb{Z}^d$, such that

$$\mathbb{P}_{\lambda,\delta}\big((x, s) \leftrightarrow (y, t)\big) \leq \exp\big[-m\delta_q(x, s; y, t)\big] \quad \text{if} \quad \delta_q(x, s; y, t) \geq D_x,$$

for $(x, s), (y, t) \in \mathbb{Z}^d \times \mathbb{R}$.

This version of the theorem of Klein can be found with explanation in [118]. It is proved by a so-called multiscale analysis.

The contact process also may inhabit a random environment in which the infection rates $\lambda_{x,y}$ and cure rates δ_x are independent random variables. Very much the same questions may be posed as for disordered percolation. There is in addition a variety of models of physics and applied probability for which the natural random environment is exactly of the above type. A brief survey of directed models with long-range dependence may be found with references in [111].

6.7 Exercises

6.1 Find $\alpha < 1$ such that the critical probability of oriented site percolation on $\mathbb{L}^2$ satisfies $\vec{p}_c^{\,\text{site}} \leq \alpha$.

6.2 Let $d \geq 2$, and let $\Pi : \mathbb{Z}^d \to \mathbb{Z}$ be given by

$$\Pi(x_1, x_2, \ldots, x_d) = \sum_{i=1}^{d} x_i.$$

Let $(A_t : t \geq 0)$ denote a contact process on $\mathbb{Z}^d$ with parameter λ and starting at the origin. Show that A may be coupled with a contact process C on $\mathbb{Z}$ with parameter λd and starting at the origin, in such a way that $\Pi(A_t) \supseteq C_t$ for all t.

Deduce that the critical point $\lambda_c(d)$ of the contact model on $\mathbb{L}^d$ satisfies $\lambda_c(d) \leq d^{-1}\lambda_c(1)$.

6.3 [37] Consider unoriented space–time percolation on $\mathbb{Z} \times \mathbb{R}$, with bridges at rate λ and cuts at rate δ. By adapting the corresponding argument for bond percolation on $\mathbb{L}^2$, or otherwise, show that the percolation probability $\theta(\lambda, \delta)$ satisfies $\theta(\lambda, \lambda) = 0$ for $\lambda > 0$.

7

Gibbs states

Brook's theorem states that a positive probability measure on a finite product may be decomposed into factors indexed by the cliques of its dependency graph. Closely related to this is the well known fact that a positive measure is a spatial Markov field on a graph G if and only if it is a Gibbs state. The Ising and Potts models are introduced, and the n-vector model is mentioned.

7.1 Dependency graphs

Let $X = (X_1, X_2, \ldots, X_n)$ be a family of random variables on a given probability space. For $i, j \in V = \{1, 2, \ldots, n\}$ with $i \neq j$, we write $i \perp j$ if: X_i and X_j are independent *conditional* on $(X_k : k \neq i, j)$. The relation $\perp$ is thus symmetric, and it gives rise to a graph G with vertex set V and edge-set $E = \{\langle i, j \rangle : i \not\perp j\}$, called the *dependency graph* of X (or of its law). We shall see that the law of X may be expressed as a product over terms corresponding to complete subgraphs of G. A complete subgraph of G is called a *clique*, and we write $\mathcal{K}$ for the set of all cliques of G. For notational simplicity later, we designate the empty subset of V to be a clique, and thus $\varnothing \in \mathcal{K}$. A clique is *maximal* if no strict superset is a clique, and we write $\mathcal{M}$ for the set of maximal cliques of G.

We assume for simplicity that the X_i take values in some countable subset S of the reals $\mathbb{R}$. The law of X gives rise to a probability mass function π on S^n given by

$$\pi(\mathbf{x}) = \mathbb{P}(X_i = x_i \text{ for } i \in V), \qquad \mathbf{x} = (x_1, x_2, \ldots, x_n) \in S^n.$$

It is easily seen by the definition of independence that $i \perp j$ if and only if π may be factorized in the form

$$\pi(\mathbf{x}) = g(x_i, U)h(x_j, U), \qquad \mathbf{x} \in S^n,$$

for some functions g and h, where $U = (x_k : k \neq i, j)$. For $K \in \mathcal{K}$ and $\mathbf{x} \in S^n$, we write $\mathbf{x}_K = (x_i : i \in K)$. We call π *positive* if $\pi(\mathbf{x}) > 0$ for all $\mathbf{x} \in S^n$.

In the following, each function f_K acts on the domain S^K.

142

7.1 Theorem [54]. *Let π be a positive probability mass function on S^n. There exist functions $f_K : S^K \to [0, \infty)$, $K \in \mathcal{M}$, such that*

$$(7.2) \qquad \pi(\mathbf{x}) = \prod_{K \in \mathcal{M}} f_K(\mathbf{x}_K), \qquad \mathbf{x} \in S^n.$$

In the simplest non-trivial example, let us assume that $i \perp j$ whenever $|i - j| \geq 2$. The maximal cliques are the pairs $\{i, i + 1\}$, and the mass function π may be expressed in the form

$$\pi(\mathbf{x}) = \prod_{i=1}^{n-1} f_i(x_i, x_{i+1}), \qquad \mathbf{x} \in S^n,$$

so that X is a Markov chain, whatever the direction of time.

Proof. We shall show that π may be expressed in the form

$$(7.3) \qquad \pi(\mathbf{x}) = \prod_{K \in \mathcal{K}} f_K(\mathbf{x}_K), \qquad \mathbf{x} \in S^n,$$

for suitable f_K. Representation (7.2) follows from (7.3) by associating each f_K with some maximal clique K' containing K as a subset.

A representation of π in the form

$$\pi(\mathbf{x}) = \prod_r f_r(\mathbf{x})$$

is said to *separate i and j* if every f_r is a constant function of either x_i or x_j, that is, no f_r depends non-trivially on both x_i and x_j. Let

$$(7.4) \qquad \pi(\mathbf{x}) = \prod_{A \in \mathcal{A}} f_A(\mathbf{x}_A)$$

be a factorization of π for some family $\mathcal{A}$ of subsets of V, and suppose that i, j satisfies: $i \perp j$, but i and j are not separated in (7.4). We shall construct from (7.4) a factorization that separates every pair r, s that is separated in (7.4), and in addition separates i, j. Continuing by iteration, we obtain a factorization that separates every pair i, j satisfying $i \perp j$, and this has the required form (7.3).

Since $i \perp j$, π may be expressed in the form

$$(7.5) \qquad \pi(\mathbf{x}) = g(x_i, U)h(x_j, U)$$

for some g, h, where $U = (x_k : j \neq i, j)$. Fix $s, t \in S$, and write $h|_t$ (respectively, $h|_{s,t}$) for the function $h(\mathbf{x})$ evaluated with $x_j = t$ (respectively, $x_i = s$, $x_j = t$). By (7.4),

$$(7.6) \qquad \pi(\mathbf{x}) = \pi(\mathbf{x})\big|_t \frac{\pi(\mathbf{x})}{\pi(\mathbf{x})\big|_t} = \left(\prod_{A \in \mathcal{A}} f_A(\mathbf{x}_A)\big|_t \right) \frac{\pi(\mathbf{x})}{\pi(\mathbf{x})\big|_t}.$$

By (7.5), the ratio

$$\frac{\pi(\mathbf{x})}{\pi(\mathbf{x})\big|_t} = \frac{h(x_j, U)}{h(t, U)}$$

is independent of x_i, so that

$$\frac{\pi(\mathbf{x})}{\pi(\mathbf{x})\big|_t} = \prod_{A \in \mathcal{A}} \frac{f_A(\mathbf{x}_A)\big|_s}{f_A(\mathbf{x}_A)\big|_{s,t}}.$$

By (7.6),

$$\pi(\mathbf{x}) = \left(\prod_{A \in \mathcal{A}} f_A(\mathbf{x}_A)\big|_t \right) \left(\prod_{A \in \mathcal{A}} \frac{f_A(\mathbf{x}_A)\big|_s}{f_A(\mathbf{x}_A)\big|_{s,t}} \right)$$

is the required representation, and the claim is proved.   □

7.2 Markov and Gibbs random fields

Let $G = (V, E)$ be a finite graph, taken for simplicity without loops or multiple edges. Within statistics and statistical mechanics, there has been a great deal of interest in probability measures having a type of 'spatial Markov property' given in terms of the neighbour relation of G. We shall restrict ourselves here to measures on the sample space $\Sigma = \{0, 1\}^V$, while noting that the following results may be extended without material difficulty to a larger product S^V, where S is finite or countably infinite.

The vector $\sigma \in \Sigma$ may be placed in one–one correspondence with the subset $\eta(\sigma) = \{v \in V : \sigma_v = 1\}$ of V, and we shall use this correspondence freely. For any $W \subseteq V$, we define the *external boundary*

$$\Delta W = \{v \in V : v \notin W, \, v \sim w \text{ for some } w \in W\}.$$

For $s = (s_v : v \in V) \in \Sigma$, we write s_W for the sub-vector $(s_w : w \in W)$. We refer to the configuration of vertices in W as the 'state' of W.

7.7 Definition. A probability measure π on Σ is said to be *positive* if $\pi(\sigma) > 0$ for all $\sigma \in \Sigma$. It is called a *Markov (random) field* if it is positive and: for all $W \subseteq V$, conditional on the state of $V \setminus W$, the law of the state of W depends only on the state of ΔW. That is, π satisfies the *global Markov property*

$$(7.8) \quad \pi\big(\sigma_W = s_W \,\big|\, \sigma_{V \setminus W} = s_{V \setminus W}\big) = \pi\big(\sigma_W = s_W \,\big|\, \sigma_{\Delta W} = s_{\Delta W}\big),$$

for all $s \in \Sigma$, and $W \subseteq V$.

The key result about such measures is their representation in terms of a 'potential function' ϕ, in a form known as a Gibbs random field (or sometimes 'Gibbs state'). Recall the set $\mathcal{K}$ of cliques of the graph G, and write 2^V for the set of all subsets (or 'power set') of V.

7.9 Definition. A probability measure π on Σ is called a *Gibbs (random)*
field if there exists a 'potential' function $\phi : 2^V \to \mathbb{R}$, satisfying $\phi_C = 0$ if
$C \notin \mathcal{K}$, such that

(7.10) $$\pi(B) = \exp\left(\sum_{K \subseteq B} \phi_K\right), \qquad B \subseteq V.$$

We allow the empty set in the above summation, so that $\log \pi(\varnothing) = \phi_\varnothing$.

Condition (7.10) has been chosen for combinatorial simplicity. It is
not the physicists' preferred definition of a Gibbs state. Let us define a
Gibbs state as a probability measure π on Σ such that there exist functions
$f_K : \{0, 1\}^K \to \mathbb{R}$, $K \in \mathcal{K}$, with

(7.11) $$\pi(\sigma) = \exp\left(\sum_{K \in \mathcal{K}} f_K(\sigma_K)\right), \qquad \sigma \in \Sigma.$$

It is immediate that π satisfies (7.10) for some ϕ whenever it satisfies (7.11).
The converse holds also, and is left for Exercise 7.1.

Gibbs fields are thus named after Josiah Willard Gibbs, whose volume
[95] made available the foundations of statistical mechanics. A simplistic
motivation for the form of (7.10) is as follows. Suppose that each state σ
has an energy E_σ, and a probability $\pi(\sigma)$. We constrain the average energy
$E = \sum_\sigma E_\sigma \pi(\sigma)$ to be fixed, and we maximize the entropy

$$\eta(\pi) = -\sum_{\sigma \in \Sigma} \pi(\sigma) \log_2 \pi(\sigma).$$

With the aid of a Lagrange multiplier β, we find that

$$\pi(\sigma) \propto e^{-\beta E_\sigma}, \qquad \sigma \in \Sigma.$$

The theory of thermodynamics leads to the expression $\beta = 1/(kT)$ where
k is Boltzmann's constant and T is (absolute) temperature. Formula (7.10)
arises when the energy E_σ may be expressed as the sum of the energies of
the sub-systems indexed by cliques.

7.12 Theorem. *A positive probability measure π on Σ is a Markov random*
field if and only if it is a Gibbs random field. The potential function ϕ
corresponding to the Markov field π is given by

$$\phi_K = \sum_{L \subseteq K} (-1)^{|K \setminus L|} \log \pi(L), \qquad K \in \mathcal{K}.$$

A positive probability measure π is said to have the *local Markov property*
if it satisfies the global property (7.8) for all *singleton* sets W and all $s \in \Sigma$.
The global property evidently implies the local property, and it turns out that
the two properties are equivalent. For notational convenience, we denote a
singleton set $\{w\}$ as w.

7.13 Proposition. *Let π be a positive probability measure on Σ. The following three statements are equivalent:*

(a) *π satisfies the global Markov property,*

(b) *π satisfies the local Markov property,*

(c) *for all $A \subseteq V$ and any pair $u, v \in V$ with $u \notin A$, $v \in A$ and $u \nsim v$,*

$$(7.14) \qquad \frac{\pi(A \cup u)}{\pi(A)} = \frac{\pi(A \cup u \setminus v)}{\pi(A \setminus v)}.$$

Proof. First, assume (a), so that (b) holds trivially. Let $u \notin A$, $v \in A$, and $u \nsim v$. Applying (7.8) with $W = \{u\}$ and, for $w \neq u$, $s_w = 1$ if and only if $w \in A$, we find that

$$(7.15)$$

$$\begin{aligned}
\frac{\pi(A \cup u)}{\pi(A) + \pi(A \cup u)} &= \pi(\sigma_u = 1 \mid \sigma_{V \setminus u} = A) \\
&= \pi(\sigma_u = 1 \mid \sigma_{\Delta u} = A \cap \Delta u) \\
&= \pi(\sigma_u = 1 \mid \sigma_{V \setminus u} = A \setminus v) \qquad \text{since } v \notin \Delta u \\
&= \frac{\pi(A \cup u \setminus v)}{\pi(A \setminus v) + \pi(A \cup u \setminus v)}.
\end{aligned}$$

Equation (7.15) is equivalent to (7.14), whence (b) and (c) are equivalent under (a).

It remains to show that the local property implies the global property. The proof requires a short calculation, and may be done either by Theorem 7.1 or within the proof of Theorem 7.12. We follow the first route here. Assume that π is positive and satisfies the local Markov property. Then $u \perp v$ for all $u, v \in V$ with $u \nsim v$. By Theorem 7.1, there exist functions f_K, $K \in \mathcal{M}$, such that

$$(7.16) \qquad \pi(A) = \prod_{K \in \mathcal{M}} f_K(A \cap K), \qquad A \subseteq V.$$

Let $W \subseteq V$. By (7.16), for $A \subseteq W$ and $C \subseteq V \setminus W$,

$$\pi(\sigma_W = A \mid \sigma_{V \setminus W} = C) = \frac{\prod_{K \in \mathcal{M}} f_K((A \cup C) \cap K)}{\sum_{B \subseteq W} \prod_{K \in \mathcal{M}} f_K((B \cup C) \cap K)}.$$

Any clique K with $K \cap W = \varnothing$ makes the same contribution $f_K(C \cap K)$ to both numerator and denominator, and may be cancelled. The remaining cliques are subsets of $\widehat{W} = W \cup \Delta W$, so that

$$\pi(\sigma_W = A \mid \sigma_{V \setminus W} = C) = \frac{\prod_{K \in \mathcal{M}, \, K \subseteq \widehat{W}} f_K((A \cup C) \cap K)}{\sum_{B \subseteq W} \prod_{K \in \mathcal{M}, \, K \subseteq \widehat{W}} f_K((B \cup C) \cap K)}.$$

The right side does not depend on $\sigma_{V \setminus \widehat{W}}$, whence

$$\pi(\sigma_W = A \mid \sigma_{V \setminus W} = C) = \pi(\sigma_W = A \mid \sigma_{\Delta W} = C \cap \Delta W)$$

as required for the global Markov property. $\qquad\square$

Proof of Theorem 7.12. Assume first that π is a positive Markov field, and let

(7.17) $$\phi_C = \sum_{L \subseteq C} (-1)^{|C \setminus L|} \log \pi(L), \qquad C \subseteq V.$$

By the inclusion–exclusion principle,

$$\log \pi(B) = \sum_{C \subseteq B} \phi_C, \qquad B \subseteq V,$$

and we need only show that $\phi_C = 0$ for $C \notin \mathcal{K}$. Suppose $u, v \in C$ and $u \nsim v$. By (7.17),

$$\phi_C = \sum_{L \subseteq C \setminus \{u,v\}} (-1)^{|C \setminus L|} \log \left(\frac{\pi(L \cup u \cup v)}{\pi(L \cup u)} \bigg/ \frac{\pi(L \cup v)}{\pi(L)} \right),$$

which equals zero by the local Markov property and Proposition 7.13. Therefore, π is a Gibbs field with potential function ϕ.

Conversely, suppose that π is a Gibbs field with potential function ϕ. Evidently, π is positive. Let $A \subseteq V$, and $u \notin A$, $v \in A$ with $u \nsim v$. By (7.10),

$$
\begin{aligned}
\log \left(\frac{\pi(A \cup u)}{\pi(A)} \right) &= \sum_{\substack{K \subseteq A \cup u, \, u \in K \\ K \in \mathcal{K}}} \phi_K \\
&= \sum_{\substack{K \subseteq A \cup u \setminus v, \, u \in K \\ K \in \mathcal{K}}} \phi_K \qquad \text{since } u \nsim v \text{ and } K \in \mathcal{K} \\
&= \log \left(\frac{\pi(A \cup u \setminus v)}{\pi(A \setminus v)} \right).
\end{aligned}
$$

The claim follows by Proposition 7.13. $\qquad\square$

We close this section with some notes on the history of the equivalence of Markov and Gibbs random fields. This may be derived from Brook's theorem, Theorem 7.1, but it is perhaps more informative to prove it directly as above via the inclusion–exclusion principle. It is normally attributed to Hammersley and Clifford, and an account was circulated (with a more complicated formulation and proof) in an unpublished note of 1971, [129] (see also [68]). Versions of Theorem 7.12 may be found in the later work of several authors, and the above proof is taken essentially from [103]. The assumption of positivity is important, and complications arise for non-positive measures, see [191] and Exercise 7.2.

For applications of the Gibbs/Markov equivalence in statistics, see, for example, [159].

7.3 Ising and Potts models

In a famous experiment, a piece of iron is exposed to a magnetic field. The field is increased from zero to a maximum, and then diminished to zero. If the temperature is sufficiently low, the iron retains some residual magnetization, otherwise it does not. There is a critical temperature for this phenomenon, often named the *Curie point* after Pierre Curie, who reported this discovery in his 1895 thesis. The famous (Lenz–)Ising model for such ferromagnetism, [142], may be summarized as follows. Let particles be positioned at the points of some lattice in Euclidean space. Each particle may be in either of two states, representing the physical states of 'spin-up' and 'spin-down'. Spin-values are chosen at random according to a Gibbs state governed by interactions between neighbouring particles, and given in the following way.

Let $G = (V, E)$ be a finite graph representing part of the lattice. Each vertex $x \in V$ is considered as being occupied by a particle that has a random spin. Spins are assumed to come in two basic types ('up' and 'down'), and thus we take the set $\Sigma = \{-1, +1\}^V$ as the sample space. The appropriate probability mass function $\lambda_{\beta,J,h}$ on Σ has three parameters satisfying $\beta, J \in [0, \infty)$ and $h \in \mathbb{R}$, and is given by

$$(7.18) \qquad \lambda_{\beta,J,h}(\sigma) = \frac{1}{Z_I} e^{-\beta H(\sigma)}, \qquad \sigma \in \Sigma,$$

where the 'Hamiltonian' $H : \Sigma \to \mathbb{R}$ and the 'partition function' Z_I are given by

$$(7.19) \quad H(\sigma) = -J \sum_{e=\langle x,y \rangle \in E} \sigma_x \sigma_y - h \sum_{x \in V} \sigma_x, \qquad Z_I = \sum_{\sigma \in \Sigma} e^{-\beta H(\sigma)}.$$

The physical interpretation of β is as the reciprocal $1/T$ of temperature, of J as the strength of interaction between neighbours, and of h as the external magnetic field. We shall consider here only the case of zero external-field, and we assume henceforth that $h = 0$. Since J is assumed non-negative, the measure $\lambda_{\beta,J,0}$ is larger for smaller $H(\sigma)$. Thus, it places greater weight on configurations having many neighbour-pairs with like spins, and for this reason it is called 'ferromagnetic'. When $J < 0$, it is called 'antiferromagnetic'.

Each edge has equal interaction strength J in the above formulation. Since β and J occur only as a product βJ, the measure $\lambda_{\beta, J, 0}$ has effectively only a single parameter βJ. In a more complicated measure not studied here, different edges e are permitted to have different interaction strengths J_e. In the meantime we shall set $J = 1$, and write $\lambda_\beta = \lambda_{\beta, 1, 0}$

Whereas the Ising model permits only two possible spin-values at each vertex, the so-called (Domb–)Potts model [202] has a general number $q \geq 2$, and is governed by the following probability measure.

Let q be an integer satisfying $q \geq 2$, and take as sample space the set of vectors $\Sigma = \{1, 2, \ldots, q\}^V$. Thus each vertex of G may be in any of q states. For an edge $e = \langle x, y \rangle$ and a configuration $\sigma = (\sigma_x : x \in V) \in \Sigma$, we write $\delta_e(\sigma) = \delta_{\sigma_x, \sigma_y}$, where $\delta_{i,j}$ is the Kronecker delta. The relevant probability measure is given by

$$(7.20) \qquad \pi_{\beta, q}(\sigma) = \frac{1}{Z_P} e^{-\beta H'(\sigma)}, \qquad \sigma \in \Sigma,$$

where $Z_P = Z_P(\beta, q)$ is the appropriate partition function (or normalizing constant) and the Hamiltonian H' is given by

$$(7.21) \qquad H'(\sigma) = - \sum_{e = \langle x, y \rangle \in E} \delta_e(\sigma).$$

In the special case $q = 2$,

$$(7.22) \qquad \delta_{\sigma_1, \sigma_2} = \tfrac{1}{2}(1 + \sigma_1 \sigma_2), \qquad \sigma_1, \sigma_2 \in \{-1, +1\},$$

It is easy to see in this case that the ensuing Potts model is simply the Ising model with an adjusted value of β, in that $\pi_{\beta, 2}$ is the measure obtained from $\lambda_{\beta/2}$ by re-labelling the local states.

We mention one further generalization of the Ising model, namely the so-called n-vector or O(n) model. Let $n \in \{1, 2, \ldots\}$ and let S^{n-1} be the set of vectors of $\mathbb{R}^n$ with unit length, that is, the $(n-1)$-sphere. A 'model' is said to have O(n) symmetry if its Hamiltonian is invariant under the operation on S^{n-1} of $n \times n$ orthonormal matrices. One such model is the n-vector model on $G = (V, E)$, with Hamiltonian

$$H_n(\mathbf{s}) = - \sum_{e = \langle x, y \rangle \in E} \mathbf{s}_x \cdot \mathbf{s}_y, \qquad \mathbf{s} = (\mathbf{s}_v : v \in V) \in (S^{n-1})^V,$$

where $\mathbf{s}_x \cdot \mathbf{s}_y$ denotes the scalar product. When $n = 1$, this is simply the Ising model. It is called the X/Y model when $n = 2$, and the Heisenberg model when $n = 3$.

The Ising and Potts models have very rich theories, and are amongst the most intensively studied of models of statistical mechanics. In 'classical' work, they are studied via cluster expansions and correlation inequalities. The so-called 'random-cluster model', developed by Fortuin and Kasteleyn around 1960, provides a single framework incorporating the percolation, Ising, and Potts models, as well as electrical networks, uniform spanning trees, and forests. It enables a representation of the two-point correlation function of a Potts model as a connection probability of an appropriate model of stochastic geometry, and this in turn allows the use of geometrical techniques already refined in the case of percolation. The random-cluster model is defined and described in Chapter 8, see also [109].

The $q = 2$ Potts model is essentially the Ising model, and special features of the number 2 allow a special analysis for the Ising model not yet replicated for general Potts models. This method is termed the 'random-current representation', and it has been especially fruitful in the study of the phase transition of the Ising model on $\mathbb{L}^d$. See [3, 7, 10] and [109, Chap. 9].

7.4 Exercises

7.1 Let $G = (V, E)$ be a finite graph, and let π be a probability measure on the power set $\Sigma = \{0, 1\}^V$. A configuration $\sigma \in \Sigma$ is identified with the subset of V on which it takes the value 1, that is, with the set $\eta(\sigma) = \{v \in V : \sigma_v = 1\}$. Show that

$$\pi(B) = \exp\left(\sum_{K \subseteq B} \phi_K\right), \qquad B \subseteq V,$$

for some function ϕ acting on the set $\mathcal{K}$ of cliques of G, if and only if

$$\pi(\sigma) = \exp\left(\sum_{K \in \mathcal{K}} f_K(\sigma_K)\right), \qquad \sigma \in \Sigma,$$

for some functions $f_K : \{0, 1\}^K \to \mathbb{R}$, with K ranging over $\mathcal{K}$. Recall the notation $\sigma_K = (\sigma_v : v \in K)$.

7.2 [191] Investigate the Gibbs/Markov equivalence for probability measures that have zeroes. It may be useful to consider the example illustrated in Figure 7.1. The graph $G = (V, E)$ is a 4-cycle, and the local state space is $\{0, 1\}$. Each of the eight configurations of the figure has probability $\frac{1}{8}$, and the other eight configurations have probability 0. Show that this measure μ satisfies the local Markov property, but cannot be written in the form

$$\mu(B) = \prod_{K \subseteq B} f(K), \qquad B \subseteq V,$$

for some f satisfying $f(K) = 1$ if $K \notin \mathcal{K}$, the set of cliques.

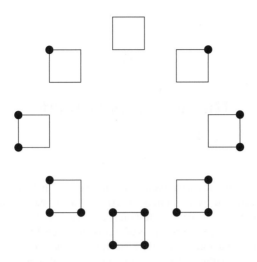

Figure 7.1. Each vertex of the 4-cycle may be in either of the two states 0 and 1. The marked vertices have state 1, and the unmarked vertices have state 0. Each of the above eight configurations has probability $\frac{1}{8}$, and the other eight configurations have probability 0.

7.3 *Ising model with external field.* Let $G = (V, E)$ be a finite graph, and let λ be the probability measure on $\Sigma = \{-1, +1\}^V$ satisfying

$$\lambda(\sigma) \propto \exp\left(h \sum_{v \in V} \sigma_v + \beta \sum_{e = \langle u, v \rangle} \sigma_u \sigma_v \right), \quad \sigma \in \Sigma,$$

where $\beta > 0$. Thinking of Σ as a partially ordered set (where $\sigma \le \sigma'$ if and only if $\sigma_v \le \sigma_v'$ for all $v \in V$), show that:

(a) λ satisfies the FKG lattice condition, and hence is positively associated,
(b) for $v \in V$, $\lambda(\cdot \mid \sigma_v = -1) \le_{\mathrm{st}} \lambda \le_{\mathrm{st}} \lambda(\cdot \mid \sigma_v = +1)$.

8

Random-cluster model

The basic properties of the model are summarized, and its relationship
to the Ising and Potts models described. The phase transition is
defined in terms of the infinite-volume measures. After an account
of a number of areas meritorious of further research, there is a section
devoted to planar duality and the conjectured value of the critical point
on the square lattice. The random-cluster model is linked in more
than one way to the study of a random even subgraph of a graph.

8.1 The random-cluster and Ising/Potts models

Let $G = (V, E)$ be a finite graph, and write $\Omega = \{0, 1\}^E$. For $\omega \in \Omega$, we
write $\eta(\omega) = \{e \in E : \omega(e) = 1\}$ for the set of open edges, and $k(\omega)$ for
the number of connected components[1], or 'open clusters', of the subgraph
$(V, \eta(\omega))$. The *random-cluster measure* on Ω, with parameters $p \in [0, 1]$,
$q \in (0, \infty)$ is the probability measure given by

$$(8.1) \qquad \phi_{p,q}(\omega) = \frac{1}{Z} \left\{ \prod_{e \in E} p^{\omega(e)} (1 - p)^{1 - \omega(e)} \right\} q^{k(\omega)}, \qquad \omega \in \Omega,$$

where $Z = Z_{G,p,q}$ is the normalizing constant.

This measure was introduced by Fortuin and Kasteleyn in a series of
papers dated around 1970. They sought a unification of the theory of electrical networks, percolation, Ising, and Potts models, and were motivated by
the observation that each of these systems satisfies a certain series/parallel
law. Percolation is evidently retrieved by setting $q = 1$, and it turns out
that electrical networks arise via the UST limit obtained on taking the limit
$p, q \to 0$ in such a way that $q/p \to 0$. The relationship to Ising/Potts
models is more complex in that it involves a transformation of measures
described next. In brief, connection probabilities for the random-cluster
measure correspond to correlations for ferromagnetic Ising/Potts models,
and this allows a geometrical interpretation of their correlation structure.

---

[1]It is important to include isolated vertices in this count.

A fuller account of the random-cluster model and its history and associations may be found in [109]. When the emphasis is upon its connection to Ising/Potts models, the random-cluster model is often called the 'FK representation'.

In the remainder of this section, we summarize the relationship between a Potts model on $G = (V, E)$ with an integer number q of local states, and the random-cluster measure $\phi_{p,q}$. As configuration space for the Potts model, we take $\Sigma = \{1, 2, \ldots, q\}^V$. Let F be the subset of the product space $\Sigma \times \Omega$ containing all pairs (σ, ω) such that: for every edge $e = \langle x, y \rangle \in E$, if $\omega(e) = 1$, then $\sigma_x = \sigma_y$. That is, F contains all pairs (σ, ω) such that σ is constant on each cluster of ω.

Let $\phi_p = \phi_{p,1}$ be product measure on ω with density p, and let μ be the probability measure on $\Sigma \times \Omega$ given by

$$(8.2) \qquad \mu(\sigma, \omega) \propto \phi_p(\omega) 1_F(\sigma, \omega), \qquad (\sigma, \omega) \in \Sigma \times \Omega,$$

where 1_F is the indicator function of F.

Four calculations are now required, in order to determine the two marginal measures of μ and the two conditional measures. It turns out that the two marginals are exactly the q-state Potts measure on Σ (with suitable pair-interaction) and the random-cluster measure $\phi_{p,q}$.

Marginal on Σ. When we sum $\mu(\sigma, \omega)$ over $\omega \in \Omega$, we have a free choice except in that $\omega(e) = 0$ whenever $\sigma_x \neq \sigma_y$. That is, if $\sigma_x = \sigma_y$, there is no constraint on the local state $\omega(e)$ of the edge $e = \langle x, y \rangle$; the sum for this edge is simply $p + (1 - p) = 1$. We are left with edges e with $\sigma_x \neq \sigma_y$, and therefore

$$(8.3) \qquad \mu(\sigma, \cdot) := \sum_{\omega \in \Omega} \mu(\sigma, \omega) \propto \prod_{e \in E} (1 - p)^{1 - \delta_e(\sigma)},$$

where $\delta_e(\sigma)$ is the Kronecker delta

$$(8.4) \qquad \delta_e(\sigma) = \delta_{\sigma_x, \sigma_y}, \qquad e = \langle x, y \rangle \in E.$$

Otherwise expressed,

$$\mu(\sigma, \cdot) \propto \exp\left\{ \beta \sum_{e \in E} \delta_e(\sigma) \right\}, \qquad \sigma \in \Sigma,$$

where

$$(8.5) \qquad p = 1 - e^{-\beta}.$$

This is the Potts measure $\pi_{\beta,q}$ of (7.20). Note that $\beta \geq 0$, which is to say that the model is ferromagnetic.

Marginal on Ω. For given ω, the constraint on σ is that it be constant on open clusters. There are $q^{k(\omega)}$ such spin configurations, and $\mu(\sigma, \omega)$ is constant on this set. Therefore,

$$\mu(\cdot, \omega) := \sum_{\sigma \in \Sigma} \mu(\sigma, \omega) \propto \left\{ \prod_{e \in E} p^{\omega(e)}(1-p)^{1-\omega(e)} \right\} q^{k(\omega)}$$

$$\propto \phi_{p,q}(\omega), \qquad \omega \in \Omega.$$

This is the random-cluster measure of (8.1).

The conditional measures. It is a routine exercise to verify the following. Given ω, the conditional measure on Σ is obtained by putting (uniformly) random spins on entire clusters of ω, constant on given clusters, and independent between clusters. Given σ, the conditional measure on Ω is obtained by setting $\omega(e) = 0$ if $\delta_e(\sigma) = 0$, and otherwise $\omega(e) = 1$ with probability p (independently of other edges).

The 'two-point correlation function' of the Potts measure $\pi_{\beta,q}$ on $G = (V, E)$ is the function $\tau_{\beta,q}$ given by

$$\tau_{\beta,q}(x, y) = \pi_{\beta,q}(\sigma_x = \sigma_y) - \frac{1}{q}, \qquad x, y \in V.$$

The 'two-point connectivity function' of the random-cluster measure $\phi_{p,q}$ is the probability $\phi_{p,q}(x \leftrightarrow y)$ of an open path from x to y. It turns out that these 'two-point functions' are (except for a constant factor) the same.

8.6 Theorem [148]. *For $q \in \{2, 3, \ldots\}$, $\beta \geq 0$, and $p = 1 - e^{-\beta}$,*

$$\tau_{\beta,q}(x, y) = (1 - q^{-1})\phi_{p,q}(x \leftrightarrow y).$$

Proof. We work with the conditional measure $\mu(\sigma \mid \omega)$ thus:

$$\tau_{\beta,q}(x, y) = \sum_{\sigma,\omega} \left[1_{\{\sigma_x=\sigma_y\}}(\sigma) - q^{-1} \right] \mu(\sigma, \omega)$$

$$= \sum_{\omega} \phi_{p,q}(\omega) \sum_{\sigma} \mu(\sigma \mid \omega) \left[1_{\{\sigma_x=\sigma_y\}}(\sigma) - q^{-1} \right]$$

$$= \sum_{\omega} \phi_{p,q}(\omega) \left[(1 - q^{-1}) 1_{\{x \leftrightarrow y\}}(\omega) + 0 \cdot 1_{\{x \nleftrightarrow y\}}(\omega) \right]$$

$$= (1 - q^{-1})\phi_{p,q}(x \leftrightarrow y),$$

and the claim is proved. $\qquad\qquad\qquad\qquad\qquad\qquad\qquad\qquad\qquad\square$

8.2 Basic properties

We list some of the fundamental properties of random-cluster measures in this section.

8.7 Theorem. *The measure $\phi_{p,q}$ satisfies the FKG lattice condition if $q \geq 1$, and is thus positively associated.*

Proof. If $p = 0, 1$, the conclusion is obvious. Assume $0 < p < 1$, and check the FKG lattice condition (4.12), which amounts to the assertion that

$$k(\omega \vee \omega') + k(\omega \wedge \omega') \geq k(\omega) + k(\omega'), \qquad \omega, \omega' \in \Omega.$$

This is left as a graph-theoretic exercise for the reader. □

8.8 Theorem (Comparison inequalities) [89]. *We have that*

(8.9) $\quad \phi_{p',q'} \leq_{st} \phi_{p,q} \quad$ *if* $p' \leq p$, $q' \geq q$, $q' \geq 1$,

(8.10) $\phi_{p',q'} \geq_{st} \phi_{p,q} \quad$ *if* $\dfrac{p'}{q'(1-p')} \geq \dfrac{p}{q(1-p)}$, $q' \geq q$, $q' \geq 1$.

Proof. This follows by the Holley inequality, Theorem 4.4, on checking condition (4.5). □

In the next theorem, the role of the graph G is emphasized in the use of the notation $\phi_{G,p,q}$. The graph $G \backslash e$ (respectively, $G.e$) is obtained from G by deleting (respectively, contracting) the edge e.

8.11 Theorem [89]. *Let $e \in E$.*

   (a) *Conditional on $\omega(e) = 0$, the measure obtained from $\phi_{G,p,q}$ is $\phi_{G \backslash e, p, q}$.*

   (b) *Conditional on $\omega(e) = 1$, the measure obtained from $\phi_{G,p,q}$ is $\phi_{G.e, p, q}$.*

Proof. This is an elementary calculation of conditional probabilities. □

In the majority of the theory of random-cluster measures, we assume that $q \geq 1$, since then we may use positive correlations and comparisons. The case $q < 1$ is slightly mysterious. It is easy to check that random-cluster measures do not generally satisfy the FKG lattice condition when $q < 1$, and indeed that they are not positively associated (see Exercise 8.2). It is considered possible, even likely, that $\phi_{p,q}$ satisfies a property of negative association when $q < 1$, and we return to this in Section 8.4.

8.3 Infinite-volume limits and phase transition

Recall the cubic lattice $\mathbb{L}^d = (\mathbb{Z}^d, \mathbb{E}^d)$. We cannot define a random-cluster measure directly on $\mathbb{L}^d$, since it is infinite. There are two possible ways to proceed. Assume $q \geq 1$.

Let $d \geq 2$, and $\Omega = \{0, 1\}^{\mathbb{E}^d}$. The appropriate σ-field of Ω is the σ-field $\mathcal{F}$ generated by the finite-dimensional sets. Let Λ be a finite box in $\mathbb{Z}^d$. For $b \in \{0, 1\}$, define

$$\Omega_\Lambda^b = \{\omega \in \Omega : \omega(e) = b \text{ for } e \notin \mathbb{E}_\Lambda\},$$

where $\mathbb{E}_A$ is the set of edges of $\mathbb{L}^d$ joining pairs of vertices belonging to A. Each of the two values of b corresponds to a certain 'boundary condition' on Λ, and we shall be interested in the effect of these boundary conditions in the infinite-volume limit.

On Ω_Λ^b, we define a random-cluster measure $\phi_{\Lambda,p,q}^b$ as follows. For $p \in [0, 1]$ and $q \in (0, \infty)$, let
(8.12)

$$\phi_{\Lambda,p,q}^b(\omega) = \frac{1}{Z_{\Lambda,p,q}^b} \left\{ \prod_{e \in \mathbb{E}_\Lambda} p^{\omega(e)}(1 - p)^{1-\omega(e)} \right\} q^{k(\omega,\Lambda)}, \quad \omega \in \Omega_\Lambda^b,$$

where $k(\omega, \Lambda)$ is the number of clusters of $(\mathbb{Z}^d, \eta(\omega))$ that intersect Λ. Here, as before, $\eta(\omega) = \{e \in \mathbb{E}^d : \omega(e) = 1\}$ is the set of open edges. The boundary condition $b = 0$ (respectively, $b = 1$) is sometimes termed 'free' (respectively, 'wired').

8.13 Theorem [104]. *Let $q \geq 1$. The weak limits*

$$\phi_{p,q}^b = \lim_{\Lambda \to \mathbb{Z}^d} \phi_{\Lambda,p,q}^b, \qquad b = 0, 1,$$

exist, and are translation-invariant and ergodic.

The infinite-volume limit is called the 'thermodynamic limit' in physics.

Proof. Let A be an increasing cylinder event defined in terms of the edges lying in some finite set S. If $\Lambda \subseteq \Lambda'$ and Λ includes the 'base' S of the cylinder A,

$$\phi_{\Lambda,p,q}^1(A) = \phi_{\Lambda',p,q}^1(A \mid \text{all edges in } \mathbb{E}_{\Lambda'\backslash\Lambda} \text{ are open}) \geq \phi_{\Lambda',p,q}^1(A),$$

where we have used Theorem 8.11 and the FKG inequality. Therefore, the limit $\lim_{\Lambda \to \mathbb{Z}^d} \phi_{\Lambda,p,q}^1(A)$ exists by monotonicity. Since $\mathcal{F}$ is generated by such events A, the weak limit $\phi_{p,q}^1$ exists. A similar argument is valid with the inequality reversed when $b = 0$.

The translation-invariance of the $\phi_{p,q}^b$ holds in very much the same way as in the proof of Theorem 2.11. The proof of ergodicity is deferred to Exercises 8.10–8.11.   □

The measures $\phi_{p,q}^0$ and $\phi_{p,q}^1$ are called 'random-cluster measures' on $\mathbb{L}^d$ with parameters p and q, and they are extremal in the following sense. We may generate ostensibly larger families of infinite-volume random-cluster measures by either of two routes. In the first, we consider measures $\phi_{\Lambda,p,q}^\xi$ on $\mathbb{E}_\Lambda$ with more general boundary conditions ξ, in order to construct a set $\mathcal{W}_{p,q}$ of 'weak-limit random-cluster measures'. The second construction uses a type of Dobrushin–Lanford–Ruelle (DLR) formalism rather than weak limits (see [104] and [109, Chap. 4]). That is, we consider measures μ on $(\Omega, \mathcal{F})$ whose measure on any box Λ, conditional on the state ξ off Λ, is the conditional random-cluster measure $\phi_{\Lambda,p,q}^\xi$. Such a μ is called a 'DLR random-cluster measure', and we write $\mathcal{R}_{p,q}$ for the set of DLR measures. The relationship between $\mathcal{W}_{p,q}$ and $\mathcal{R}_{p,q}$ is not fully understood, and we make one remark about this. Any element μ of the closed convex hull of $\mathcal{W}_{p,q}$ with the so-called '0/1-infinite-cluster property' (that is, $\mu(I \in \{0, 1\}) = 1$, where I is the number of infinite open clusters) belongs to $\mathcal{R}_{p,q}$, see [109, Sect. 4.4]. The standard way of showing the 0/1-infinite-cluster property is via the Burton–Keane argument used in the proof of Theorem 5.22. We may show, in particular, that $\phi_{p,q}^0, \phi_{p,q}^1 \in \mathcal{R}_{p,q}$.

It is not difficult to see that the measures $\phi_{p,q}^0$ and $\phi_{p,q}^1$ are extremal in the sense that

$$(8.14) \qquad \phi_{p,q}^0 \leq_{st} \phi_{p,q} \leq_{st} \phi_{p,q}^1, \qquad \phi_{p,q} \in \mathcal{W}_{p,q} \cup \mathcal{R}_{p,q},$$

whence there exists a unique random-cluster measure (in either of the above senses) if and only if $\phi_{p,q}^0 = \phi_{p,q}^1$. It is a general fact that such extremal measures are invariably ergodic, see [94, 109].

Turning to the question of phase transition, and remembering percolation, we define the *percolation probabilities*

$$(8.15) \qquad \theta^b(p, q) = \phi_{p,q}^b(0 \leftrightarrow \infty), \qquad b = 0, 1,$$

that is, the probability that 0 belongs to an infinite open cluster. The corresponding *critical values* are given by

$$(8.16) \qquad p_c^b(q) = \sup\{p : \theta^b(p, q) = 0\}, \qquad b = 0, 1.$$

Faced possibly with two (or more) distinct critical values, we present the following result.

8.17 Theorem [9, 104]. *Let $d \geq 2$ and $q \geq 1$. We have that:*

(a) $\phi_{p,q}^0 = \phi_{p,q}^1$ *if* $\theta^1(p,q) = 0$,

(b) *there exists a countable subset $\mathcal{D}_{d,q}$ of $[0, 1]$, possibly empty, such that $\phi_{p,q}^0 = \phi_{p,q}^1$ if and only if $p \notin \mathcal{D}_{d,q}$.*

It may be shown[2] that

$$(8.18) \qquad \theta^1(p,q) = \lim_{\Lambda \uparrow \mathbb{Z}^d} \phi_{\Lambda,p,q}^1(0 \leftrightarrow \partial\Lambda).$$

It is not know when the corresponding statement with $b = 0$ holds.

Sketch proof. The argument for (a) is as follows. Clearly,

$$(8.19) \qquad \theta^1(p,q) = \lim_{\Lambda \uparrow \mathbb{Z}^d} \phi_{p,q}^1(0 \leftrightarrow \partial\Lambda).$$

Suppose $\theta^1(p,q) = 0$, and consider a large box Λ with 0 in its interior. On building the clusters that intersect the boundary $\partial\Lambda$, with high probability we do not reach 0. That is, with high probability, there exists a 'cut-surface' S between 0 and $\partial\Lambda$ comprising only closed edges. By taking S to be as large as possible, the position of S may be taken to be measurable on its exterior, whence the conditional measure on the interior of S is a free random-cluster measure. Passing to the limit as $\Lambda \uparrow \mathbb{Z}^d$, we find that the free and wired measures are equal.

The argument for (b) is based on a classical method of statistical mechanics using convexity. Let $Z_{G,p,q}$ be the partition function of the random-cluster model on a graph $G = (V, E)$, and set

$$Y_{G,p,q} = (1 - p)^{-|E|} Z_{G,p,q} = \sum_{\omega \in \{0,1\}^E} e^{\pi |\eta(\omega)|} q^{k(\omega)},$$

where $\pi = \log[p/(1 - p)]$. It is easily seen that $\log Y_{G,p,q}$ is a convex function of π. By a standard method based on the negligibility of the boundary of a large box Λ compared with its volume, the limit 'pressure function'

$$\Pi(\pi, q) = \lim_{\Lambda \uparrow \mathbb{Z}^d} \left\{ \frac{1}{|\mathbb{E}_\Lambda|} \log Y_{\Lambda,p,q}^\xi \right\}$$

exists and is independent of the boundary configuration $\xi \in \Omega$. Since Π is the limit of convex functions of π, it is convex, and hence differentiable except on some countable set $\mathcal{D}$ of values of π. Furthermore, for $\pi \notin \mathcal{D}$, the derivative of $|\mathbb{E}_\Lambda|^{-1} \log Y_{\Lambda,p,q}^\xi$ converges to that of Π. The former derivative may be interpreted in terms of the edge-densities of the measures,

---

[2]Exercise 8.8.

and therefore the limits of the last are independent of ξ for any π at which $\Pi(\pi, q)$ is differentiable. Uniqueness of random-cluster measures follows by (8.14) and stochastic ordering: if μ_1, μ_2 are probability measures on $(\Omega, \mathcal{F})$ with $\mu_1 \leq_{st} \mu_2$ and satisfying

$$\mu_1(e \text{ is open}) = \mu_2(e \text{ is open}), \qquad e \in \mathbb{E},$$

then $\mu_1 = \mu_2.^3$ □

By Theorem 8.17, $\theta^0(p, q) = \theta^1(p, q)$ for $p \notin \mathcal{D}_{d,q}$, whence $p_c^0(q) = p_c^1(q)$. Henceforth we refer to the critical value as $p_c = p_c(q)$. It is a basic fact that $p_c(q)$ is non-trivial.

8.20 Theorem [9]. *If $d \geq 2$ and $q \geq 1$, then $0 < p_c(q) < 1$.*

It is an open problem to find a satisfactory definition of $p_c(q)$ for $q < 1$, although it may be shown by the comparison inequalities (Theorem 8.8) that there is no infinite cluster for $q \in (0, 1)$ and small p, and conversely there is an infinite cluster for $q \in (0, 1)$ and large p.

Proof. Let $q \geq 1$. By Theorem 8.8, $\phi^1_{p',1} \leq_{st} \phi^1_{p,q} \leq_{st} \phi_{p,1}$, where

$$p' = \frac{p}{p + q(1 - p)}.$$

We apply this inequality to the increasing event $\{0 \leftrightarrow \partial \Lambda\}$, and let $\Lambda \uparrow \mathbb{Z}^d$ to obtain via (8.18) that

$$(8.21) \qquad p_c(1) \leq p_c(q) \leq \frac{q p_c(1)}{1 + (q - 1) p_c(1)}, \qquad q \geq 1,$$

where $0 < p_c(1) < 1$ by Theorem 3.2.   □

The following is an important conjecture.

8.22 Conjecture. *There exists $Q = Q(d)$ such that:*
(a) *if $q < Q(d)$, then $\theta^1(p_c, q) = 0$ and $\mathcal{D}_{d,q} = \varnothing$,*
(b) *if $q > Q(d)$, then $\theta^1(p_c, q) > 0$ and $\mathcal{D}_{d,q} = \{p_c\}$.*

In the physical vernacular, there is conjectured a critical value of q beneath which the phase transition is continuous ('second order') and above which it is discontinuous ('first order'). Following work of Roman Kotecký and Senya Shlosman [156], it was proved in [157] that there is a first-order transition for large q, see [109, Sects 6.4, 7.5]. It is expected[4] that

$$Q(d) = \begin{cases} 4 & \text{if } d = 2, \\ 2 & \text{if } d \geq 6. \end{cases}$$

---

[3]*Exercise.* Recall Strassen's Theorem 4.2.
[4]See [25, 138, 242] for discussions of the two-dimensional case.

This may be contrasted with the best current estimate in two dimensions, namely $Q(2) \leq 25.72$, see [109, Sect. 6.4].

Finally, we review the relationship between the random-cluster and Potts phase transitions. The 'order parameter' of the Potts model is the 'magnetization' given by

$$M(\beta, q) = \lim_{\Lambda \to \mathbb{Z}^d} \left\{ \pi^1_{\Lambda,\beta}(\sigma_0 = 1) - \frac{1}{q} \right\},$$

where $\pi^1_{\Lambda,\beta}$ is the Potts measure on Λ 'with boundary condition 1'. We may think of $M(\beta, q)$ as a measure of the degree to which the boundary condition '1' is noticed at the origin after taking the infinite-volume limit. By an application of Theorem 8.6 to a suitable graph obtained from Λ,

$$\pi^1_{\Lambda,q}(\sigma_0 = 1) - \frac{1}{q} = (1 - q^{-1})\phi^1_{\Lambda,p,q}(0 \leftrightarrow \partial \Lambda),$$

where $p = 1 - e^{-\beta}$. By (8.18),

$$M(\beta, q) = (1 - q^{-1}) \lim_{\Lambda \to \mathbb{Z}^d} \phi^1_{\Lambda,p,q}(0 \leftrightarrow \partial \Lambda)$$

$$= (1 - q^{-1})\theta^1(p, q).$$

That is, $M(\beta, q)$ and $\theta^1(p, q)$ differ by the factor $1 - q^{-1}$.

8.4 Open problems

Many questions remain at least partly unanswered for the random-cluster model, and we list a few of these here. Further details may be found in [109].

I. *The case $q < 1$.* Less is known when $q < 1$ owing to the failure of the FKG inequality. A possibly optimistic conjecture is that some version of negative association holds when $q < 1$, and this might imply the existence of infinite-volume limits. Possibly the weakest conjecture is that

$$\phi_{p,q}(e \text{ and } f \text{ are open}) \leq \phi_{p,q}(e \text{ is open})\phi_{p,q}(f \text{ is open}),$$

for distinct edges e and f. It has not been ruled out that $\phi_{p,q}$ satisfies the stronger BK inequality when $q < 1$. Weak limits of $\phi_{p,q}$ as $q \downarrow 0$ have a special combinatorial structure, but even here the full picture has yet to emerge. More specifically, it is not hard to see that

$$\phi_{p,q} \Rightarrow \begin{cases} \text{UCS} & \text{if } p = \frac{1}{2}, \\ \text{UST} & \text{if } p \to 0 \text{ and } q/p \to 0, \\ \text{UF} & \text{if } p = q, \end{cases}$$

where the acronyms are the *uniform connected subgraph, uniform spanning tree*, and *uniform forest* measures encountered in Sections 2.1 and 2.4. See Theorem 2.1 and Conjecture 2.14.

We may use comparison arguments to study infinite-volume random-cluster measures for sufficiently small or large p, but there is no proof of the existence of a unique point of phase transition.

The case $q < 1$ is of more mathematical than physical interest, although the various limits as $q \to 0$ are relevant to the theory of algorithms and complexity.

Henceforth, we assume $q \geq 1$.

II. *Exponential decay.* Prove that

$$\phi_{p,q}\left(0 \leftrightarrow \partial[-n, n]^d\right) \leq e^{-\alpha n}, \qquad n \geq 1,$$

for some $\alpha = \alpha(p, q)$ satisfying $\alpha > 0$ when $p < p_c(q)$. This has been proved for sufficiently small values of p, but no proof is known (for general q and any given $d \geq 2$) right up to the critical point.

The case $q = 2$ is special, since this corresponds to the Ising model, for which the random-current representation has allowed a rich theory, see [109, Sect. 9.3]. Exponential decay is proved to hold for general d, when $q = 2$, and also for sufficiently large q (see IV below).

III. *Uniqueness of random-cluster measures.* Prove all or part of Conjecture 8.22. That is, show that $\phi_{p,q}^0 = \phi_{p,q}^1$ for $p \neq p_c(q)$; and, furthermore, that uniqueness holds when $p = p_c(q)$ if and only if q is sufficiently small.

These statements are trivial when $q = 1$, and uniqueness is proved when $q = 2$ and $p \neq p_c(2)$ using the theory of the Ising model alluded to above. The situation is curious when $q = 2$ and $p = p_c(2)$, in that uniqueness is proved so long as $d \neq 3$, see [109, Sect. 9.4].

When q is sufficiently large, it is known as in IV below that there is a unique random-cluster measure when $p \neq p_c(q)$ and a multiplicity of such measures when $p = p_c(q)$.

IV. *First/second-order phase transition.* Much interest in Potts and random-cluster measures is focussed on the fact that nature of the phase transition depends on whether q is small or large, see for example Conjecture 8.22. For small q, the singularity is expected to be continuous and of power type. For large q, there is a discontinuity in the order parameter $\theta^1(\cdot, q)$, and a 'mass gap' at the critical point (that is, when $p = p_c(q)$, the $\phi_{p,q}^0$-probability of a long path decays exponentially, while the $\phi_{p,q}^1$-probability is bounded away from 0).

Of the possible questions, we ask for a proof of the existence of a value $Q = Q(d)$ separating the second- from the first-order transition.

V. *Slab critical point.* It was important for supercritical percolation in three and more dimensions to show that percolation in $\mathbb{L}^d$ implies percolation in a sufficiently fat 'slab', see Theorem 5.17. A version of the corresponding problem for the random-cluster model is as follows. Let $q \geq 1$ and $d \geq 3$, and write $S(L, n)$ for the 'slab'

$$S(L, n) = [0, L-1] \times [-n, n]^{d-1}.$$

Let $\psi_{L,n,p,q} = \phi^0_{S(L,n),p,q}$ be the random-cluster measure on $S(L, n)$ with parameters p, q, and free boundary conditions. Let $\Pi(p, L)$ denote the property that[5]

$$\liminf_{n \to \infty} \inf_{x \in S(L,n)} \{\psi_{L,n,p,q}(0 \leftrightarrow x)\} > 0.$$

It is not hard[6] to see that $\Pi(p, L) \Rightarrow \Pi(p', L')$ if $p \leq p'$ and $L \leq L'$, and it is thus natural to define

(8.23) $\widehat{p}_c(q, L) = \inf\{p : \Pi(p, L) \text{ occurs}\}, \quad \widehat{p}_c(q) = \lim_{L \to \infty} \widehat{p}_c(q, L).$

Clearly, $p_c(q) \leq \widehat{p}_c(q) < 1$. It is believed that equality holds in that $\widehat{p}_c(q) = p_c(q)$, and it is a major open problem to prove this. A positive resolution would have implications for the exponential decay of truncated cluster-sizes, and for the existence of a Wulff crystal for all $p > p_c(q)$ and $q \geq 1$. See Figure 5.3 and [60, 61, 62].

VI. *Roughening transition.* While it is believed that there is a *unique* random-cluster measure except possibly at the critical point, there can exist a multitude of random-cluster-type measures with the striking property of non-translation-invariance. Take a box $\Lambda_n = [-n, n]^d$ in $d \geq 3$ dimensions (the following construction fails in 2 dimensions). We may think of $\partial \Lambda_n$ as comprising a northern and southern hemisphere, with the 'equator' $\{x \in \partial \Lambda_n : x_d = 0\}$ as interface. Let $\overline{\phi}_{n,p,q}$ be the random-cluster measure on Λ_n with a wired boundary condition on the northern and southern hemispheres individually and conditioned on the event D_n that no open path joins a point of the northern to a point of the southern hemisphere. By the compactness of Ω, the sequence $(\overline{\phi}_{n,p,q} : n \geq 1)$ possesses weak limits. Let $\overline{\phi}_{p,q}$ be such a weak limit.

It is a geometrical fact that, in any configuration $\omega \in D_n$, there exists an interface $I(\omega)$ separating the points joined to the northern hemisphere from

---

[5]This corrects an error in [109].
[6]Exercise 8.9.

those joined to the southern hemisphere. This interface passes around the equator, and its closest point to the origin is at some distance H_n, say. It may be shown that, for $q \geq 1$ and sufficiently large p, the laws of the H_n are tight, whence the weak limit $\overline{\phi}_{p,q}$ is not translation-invariant. Such measures are termed 'Dobrushin measures' after their discovery for the Ising model in [70].

There remain two important questions. Firstly, for $d \geq 3$ and $q \geq 1$, does there exist a value $\widetilde{p}(q)$ such that Dobrushin measures exist for $p > \widetilde{p}(q)$ and not for $p < \widetilde{p}(q)$? Secondly, for what dimensions d do Dobrushin measures exist for all $p > p_c(q)$? A fuller account may be found in [109, Chap. 7].

VII. *In two dimensions.* There remain some intriguing conjectures in the playground of the square lattice $\mathbb{L}^2$, and some of these are described in the next section.

8.5 In two dimensions

Consider the special case of the square lattice $\mathbb{L}^2$. Random-cluster measures on $\mathbb{L}^2$ have a property of self-duality that generalizes that of bond percolation. (We recall the discussion of duality after equation (3.7).) The most provocative conjecture is that the critical point equals the so-called self-dual point.

8.24 Conjecture. *For $d = 2$ and $q \geq 1$,*

$$(8.25) \qquad\qquad p_c(q) = \frac{\sqrt{q}}{1 + \sqrt{q}}.$$

This formula is proved rigorously when $q = 1$ (percolation), when $q = 2$ (Ising model), and for sufficiently large values of q (namely, $q \geq 25.72$).[7] Physicists have 'known' for some time that the self-dual point marks a discontinuous phase transition when $q > 4$.

The conjecture is motivated as follows. Let $G = (V, E)$ be a finite planar graph, and $G_d = (V_d, E_d)$ its dual graph. To each $\omega \in \Omega = \{0, 1\}^E$, there corresponds the dual configuration[8] $\omega_d \in \Omega_d = \{0, 1\}^{E_d}$, given by

$$\omega_d(e_d) = 1 - \omega(e), \qquad e \in E.$$

---

[7]*Added in proof*: Substantial progress with the above conjecture has been made recently by Vincent Beffara and Hugo Duminil-Copin [28].

[8]Note that this definition of the dual configuration differs from that used in Chapter 3 for percolation.

By drawing a picture, we may convince ourselves that every face of $(V, \eta(\omega))$ contains a unique component of $(V_d, \eta(\omega_d))$, and therefore the number $f(\omega)$ of faces (including the infinite face) of $(V, \eta(\omega))$ satisfies

$$(8.26) \qquad f(\omega) = k(\omega_d).$$

The random-cluster measure on G satisfies

$$\phi_{G,p,q}(\omega) \propto \left(\frac{p}{1-p}\right)^{|\eta(\omega)|} q^{k(\omega)}.$$

Using (8.26), Euler's formula,

$$(8.27) \qquad k(\omega) = |V| - |\eta(\omega)| + f(\omega) - 1,$$

and the fact that $|\eta(\omega)| + |\eta(\omega_d)| = |E|$, we have that

$$\phi_{G,p,q}(\omega) \propto \left(\frac{q(1-p)}{p}\right)^{|\eta(\omega_d)|} q^{k(\omega_d)},$$

which is to say that

$$(8.28) \qquad \phi_{G,p,q}(\omega) = \phi_{G_d,p_d,q}(\omega_d), \qquad \omega \in \Omega,$$

where

$$(8.29) \qquad \frac{p_d}{1-p_d} = \frac{q(1-p)}{p}.$$

The unique fixed point of the mapping $p \mapsto p_d$ is given by $p = \kappa_q$, where κ is the 'self-dual point'

$$\kappa_q = \frac{\sqrt{q}}{1+\sqrt{q}}.$$

Turning to the square lattice, let $G = \Lambda = [0, n]^2$, with dual graph $G_d = \Lambda_d$ obtained from the box $[-1, n]^2 + (\frac{1}{2}, \frac{1}{2})$ by identifying all boundary vertices. By (8.28),

$$(8.30) \qquad \phi^0_{\Lambda,p,q}(\omega) = \phi^1_{\Lambda_d,p_d,q}(\omega_d)$$

for configurations ω on Λ (and with a small 'fix' on the boundary of Λ_d). Letting $n \to \infty$, we obtain that

$$(8.31) \qquad \phi^0_{p,q}(A) = \phi^1_{p_d,q}(A_d)$$

for all cylinder events A, where $A_d = \{\omega_d : \omega \in A\}$.

The duality relation (8.31) is useful, especially if $p = p_d = \kappa_q$. In particular, the proof that $\theta(\frac{1}{2}) = 0$ for percolation (see Theorem 5.33) may be adapted to obtain

$$(8.32) \qquad \theta^0(\kappa_q, q) = 0,$$

whence

(8.33)
$$p_c(q) \geq \frac{\sqrt{q}}{1 + \sqrt{q}}, \qquad q \geq 1.$$

In order to obtain the formula of Conjecture 8.24, it would be enough to show that

$$\phi^0_{p,q}\big(0 \leftrightarrow \partial[-n, n]^2\big) \leq \frac{A}{n}, \qquad n \geq 1,$$

where $A = A(p, q) < \infty$ for $p < \kappa_q$. See [100, 109].

The case $q = 2$ is very special, because it is related to the Ising model, for which there is a rich and exact theory going back to Onsager [193]. As an illustration of this connection in action, we include a proof that the wired random-cluster measure has no infinite cluster at the self-dual point. The corresponding conclusion in believed to hold if and only if $q \leq 4$, but a full proof is elusive.

8.34 Theorem. *For $d = 2$, $\theta^1(\kappa_2, 2) = 0$.*

Proof. Of the several proofs of this statement, we outline the recent simple proof of Werner [237]. Let $q = 2$, and write $\phi^b = \phi^b_{p_{\mathrm{sd}}(q),q}$.

Let $\omega \in \Omega$ be a configuration of the random-cluster model sampled according to ϕ^0. To each open cluster of ω, we allocate the spin $+1$ with probability $\frac{1}{2}$, and -1 otherwise. Thus, spins are constant within clusters, and independent between clusters. Let σ be the resulting spin configuration, and let μ^0 be its law. We do the same with ω sampled from ϕ^1, with the difference that any infinite cluster is allocated the spin $+1$. It is not hard to see that the resulting measure μ^1 is the infinite-volume Ising measure with boundary condition $+1$.[9] The spin-space $\Sigma = \{-1, +1\}^{\mathbb{Z}^2}$ is a partially ordered set, and it may be checked using the Holley inequality[10], Theorem 4.4, and passing to an infinite-volume limit that

(8.35)
$$\mu^0 \leq_{\mathrm{st}} \mu^1.$$

We shall be interested in two notions of connectivity in $\mathbb{Z}^2$, the first of which is the usual one, denoted $\leftrightarrow$. If we add both diagonals to each face of $\mathbb{Z}^2$, we obtain a new graph with so-called $*$-connectivity relation denoted $\leftrightarrow_*$. A cycle in this new graph is called a $*$-cycle.

Each spin-vector $\sigma \in \Sigma$ amounts to a partition of $\mathbb{Z}^2$ into maximal clusters with constant spin. A cluster labelled $+1$ (respectively, -1) is called a

---

[9]This is formalized in [109, Sect. 4.6]; see also Exercise 8.16.
[10]See Exercise 7.3.

(+)-cluster (respectively, (−)-cluster). Let $N^+(\sigma)$ (respectively, $N^-(\sigma)$) be the number of infinite (+)-clusters (respectively, infinite (−)-clusters).

By (8.32), $\phi^0(0 \leftrightarrow \infty) = 0$, whence, by Exercise 8.16, μ^0 is ergodic. We may apply the Burton–Keane argument of Section 5.3 to deduce that

$$\text{either} \quad \mu^0(N^+ = 1) = 1 \quad \text{or} \quad \mu^0(N^+ = 0) = 1.$$

We may now use Zhang's argument (as in the proof of (8.32) and Theorem 5.33), and the fact that N^+ and N^- have the same law, to deduce that

$$(8.36) \qquad\qquad \mu^0(N^+ = 0) = \mu^0(N^- = 0) = 1.$$

Let A be an increasing cylinder event of Σ defined in terms of states of vertices in some box Λ. By (8.36), there are (μ^0-a.s.) no infinite (−)-clusters intersecting Λ, so that Λ lies in the interior of some ∗-cycle labelled +1. Let $\Lambda_n = [-n, n]^2$ with n large, and let D_n be the event that Λ_n contains a ∗-cycle labelled +1 with Λ in its interior. By the above, $\mu^0(D_n) \to 1$ as $n \to \infty$. The event D_n is an increasing subset of Σ, whence, by (8.35),

$$\mu^1(D_n) \to 1 \qquad \text{as } n \to \infty.$$

On the event D_n, we find the 'outermost' ∗-cycle H of Λ_n labelled +1; this cycle may be constructed explicitly via the boundaries of the (−)-clusters intersecting $\partial \Lambda_n$. Since H is outermost, the conditional measure of μ^1 (given D_n), restricted to Λ, is stochastically smaller than μ^0. On letting $n \to \infty$, we obtain $\mu^1(A) \le \mu^0(A)$, which is to say that $\mu^1 \le_{st} \mu^0$. By (8.35), $\mu^0 = \mu^1$.

By (8.36), $\mu^1(N^+ = 0) = 1$, so that $\theta^1(\kappa_2, 2) = 0$ as claimed.   □

Last, but definitely not least, we turn towards SLE, random-cluster, and Ising models. Stanislav Smirnov has recently proved the convergence of re-scaled boundaries of large clusters of the critical random-cluster model on $\mathbb{L}^2$ to $SLE_{16/3}$. The corresponding critical Ising model has spin-cluster boundaries converging to SLE_3. These results are having a major impact on our understanding of the Ising model.

This section ends with an open problem concerning the Ising model on the triangular lattice. Each Ising spin-configuration $\sigma \in \{-1, +1\}^V$ on a graph $G = (V, E)$ gives rise to a subgraph $G^\sigma = (V, E^\sigma)$ of G, where

$$(8.37) \qquad\qquad E^\sigma = \{e = \langle u, v \rangle \in E : \sigma_u = \sigma_v\}.$$

If G is planar, the boundary of any connected component of G^σ corresponds to a cycle in the dual graph G_d, and the union of all such cycles is a (random) even subgraph of G_d (see the next section).

We shall consider the Ising model on the square and triangular lattices, with inverse-temperature β satisfying $0 \leq \beta \leq \beta_c$, where β_c is the critical value. By (8.5),

$$e^{-2\beta_c} = 1 - p_c(2).$$

We begin with the square lattice $\mathbb{L}^2$, for which $p_c(2) = \sqrt{2}/(1+\sqrt{2})$. When $\beta = 0$, the model amounts to site percolation with density $\frac{1}{2}$. Since this percolation process has critical point satisfying $p_c^{\text{site}} > \frac{1}{2}$, each spin-cluster of the $\beta = 0$ Ising model is subcritical, and in particular has an exponentially decaying tail. More specifically, write $x \overset{\pm}{\longleftrightarrow} y$ if there exists a path of $\mathbb{L}^2$ from x to y with constant spin-value, and let

$$S_x = \{y \in V : x \overset{\pm}{\longleftrightarrow} y\}$$

be the spin-cluster at x, and $S = S_0$. By the above, there exists $\alpha > 0$ such that

(8.38) $$\lambda_0(|S| \geq n + 1) \leq e^{-\alpha n}, \qquad n \geq 1,$$

where λ_β denotes the infinite-volume Ising measure. It is standard (and follows from Theorem 8.17(a)) that there is a unique Gibbs state for the Ising model when $\beta < \beta_c$ (see [113, 237] for example).

The exponential decay of (8.38) extends throughout the subcritical phase in the following sense. Yasunari Higuchi [137] has proved that

(8.39) $$\lambda_\beta(|S| \geq n + 1) \leq e^{-\alpha n}, \qquad n \geq 1,$$

where $\alpha = \alpha(\beta)$ satisfies $\alpha > 0$ when $\beta < \beta_c$. There is a more recent proof of this (and more) by Rob van den Berg [34, Thm 2.4], using the sharp-threshold theorem, Theorem 4.81. Note that (8.39) implies the weaker (and known) statement that the volumes of clusters of the $q = 2$ random-cluster model on $\mathbb{L}^2$ have an exponentially decaying tail.

Inequality (8.39) fails in an interesting manner when the square lattice is replaced by the triangular lattice $\mathbb{T}$. Since $p_c^{\text{site}}(\mathbb{T}) = \frac{1}{2}$, the $\beta = 0$ Ising model is critical. In particular, the tail of $|S|$ is of power-type and, by Smirnov's theorem for percolation, the scaling limit of the spin-cluster boundaries is SLE_6. Furthermore, the process is, in the following sense, *critical* for all $\beta \in [0, \beta_c]$. Since there is a unique Gibbs state for $\beta < \beta_c$, λ_β is invariant under the interchange of spin-values $-1 \leftrightarrow +1$. Let R_n be a rhombus of the lattice with side-lengths n and axes parallel to the horizontal and one of the diagonal lattice directions, and let A_n be the event that R_n is traversed from left to right by a $+$ path (that is, a path ν satisfying $\sigma_y = +1$ for all $y \in \nu$). It is easily seen that the complement of A_n is the event that R_n

is crossed from top to bottom by a − path (see Figure 5.12 for an illustration of the analogous case of bond percolation on the square lattice). Therefore,

$$(8.40) \qquad \lambda_\beta(A_n) = \tfrac{1}{2}, \qquad 0 \le \beta < \beta_c.$$

Let S_x be the spin-cluster containing x as before, and define

$$\mathrm{rad}(S_x) = \max\{\delta(x, z) : z \in S_x\},$$

where δ denotes graph-theoretic distance. By (8.40), there exists a vertex x such that $\lambda_\beta(\mathrm{rad}(S_x) \ge n) \ge (2n)^{-1}$. By the translation-invariance of λ_β,

$$\lambda_\beta(\mathrm{rad}(S) \ge n) \ge \frac{1}{2n}, \qquad 0 \le \beta < \beta_c.$$

In conclusion, the tail of $\mathrm{rad}(S)$ is of power-type for all $\beta \in [0, \beta_c)$.

It is believed that the SLE_6 cluster-boundary limit 'propagates' from $\beta = 0$ to all values $\beta < \beta_c$. Further evidence for this may be found in [23]. When $\beta = \beta_c$, the corresponding limit is the same as that for the square lattice, namely SLE_3, see [67].

8.6 Random even graphs

A subset F of the edge-set of $G = (V, E)$ is called *even* if each vertex $v \in V$ is incident to an even number of elements of F, and we write $\mathcal{E}$ for the set of even subsets F. The subgraph (V, F) of G is *even* if F is even. It is standard that every even set F may be decomposed as an edge-disjoint union of cycles. Let $p \in [0, 1)$. The *random even subgraph* of G with parameter p is that with law

$$(8.41) \qquad \eta_p(F) = \frac{1}{Z_e} p^{|F|}(1 - p)^{|E \setminus F|}, \qquad F \in \mathcal{E},$$

where

$$Z_e = \sum_{F \in \mathcal{E}} p^{|F|}(1 - p)^{|E \setminus F|}.$$

When $p = \tfrac{1}{2}$, we talk of a *uniform* random even subgraph.

We may express η_p in the following way. Let $\phi_p = \phi_{p,1}$ be product measure with density p on $\Omega = \{0, 1\}^E$. For $\omega \in \Omega$, let $\partial\omega$ denote the set of vertices $v \in V$ that are incident to an odd number of ω-open edges. Then

$$\eta_p(F) = \frac{\phi_p(\omega_F)}{\phi_p(\partial\omega = \varnothing)}, \qquad F \in \mathcal{E},$$

where ω_F is the edge-configuration whose open set is F. In other words, ϕ_p describes the random subgraph of G obtained by randomly and independently deleting each edge with probability $1 - p$, and η_p is the law of this random subgraph conditioned on being even.

Let λ_β be the Ising measure on a graph H with inverse temperature $\beta \geq 0$, presented in the form
(8.42)

$$\lambda_\beta(\sigma) = \frac{1}{Z_{\mathrm{I}}} \exp\left(\beta \sum_{e=\langle u,v \rangle \in E} \sigma_u \sigma_v \right), \qquad \sigma = (\sigma_v : v \in V) \in \Sigma,$$

with $\Sigma = \{-1, +1\}^V$. See (7.18) and (7.20). A spin configuration σ gives rise to a subgraph $G^\sigma = (V, E^\sigma)$ of G with E^σ given in (8.37) as the set of edges whose endpoints have like spin. When G is planar, the boundary of any connected component of G^σ corresponds to a cycle of the dual graph G_{d}, and the union of all such cycles is a (random) even subgraph of G_{d}. A glance at (8.3) informs us that the law of this even graph is η_r, where

$$\frac{r}{1-r} = e^{-2\beta}.$$

Note that $r \leq \frac{1}{2}$. Thus, one way of generating a random even subgraph of a planar graph $G = (V, E)$ with parameter $r \in [0, \frac{1}{2}]$ is to take the dual of the graph G^σ with σ is chosen with law (8.42), and with $\beta = \beta(r)$ chosen suitably.

The above recipe may be cast in terms of the random-cluster model on the planar graph G. First, we sample ω according to the random-cluster measure $\phi_{p,q}$ with $p = 1 - e^{-2\beta}$ and $q = 2$. To each open cluster of ω we allocate a random spin taken uniformly from $\{-1, +1\}$. These spins are constant on clusters and independent between clusters. By the discussion of Section 8.1, the resulting spin-configuration σ has law λ_β. The boundaries of the spin-clusters may be constructed as follows from ω. Let $C_1, C_2, \ldots, C_c$ be the external boundaries of the open clusters of ω, viewed as cycles of the dual graph, and let $\xi_1, \xi_2, \ldots, \xi_c$ be independent Bernoulli random variables with parameter $\frac{1}{2}$. The sum $\sum_i \xi_i C_i$, with addition interpreted as symmetric difference, has law η_r.

It turns out that we can generate a random even subgraph of a graph G from the random-cluster model on G, for an arbitrary, possibly non-planar, graph G. We consider first the uniform case of η_p with $p = \frac{1}{2}$.

We identify the family of all spanning subgraphs of $G = (V, E)$ with the family of all subsets of E (the word 'spanning' indicates that these subgraphs have the original vertex-set V). This family can further be identified with $\Omega = \{0, 1\}^E = \mathbb{Z}_2^E$, and is thus a vector space over $\mathbb{Z}_2$; the operation $+$ of addition is component-wise addition modulo 2, which translates into taking the symmetric difference of edge-sets: $F_1 + F_2 = F_1 \triangle F_2$ for $F_1, F_2 \subseteq E$.

The family $\mathcal{E}$ of even subgraphs of G forms a subspace of the vector space $\mathbb{Z}_2^E$, since $F_1 \triangle F_2$ is even if F_1 and F_2 are even. In particular, the number of even subgraphs of G equals $2^{c(G)}$, where $c(G) = \dim(\mathcal{E})$. The quantity $c(G)$ is thus the number of independent cycles in G, and is known as the *cyclomatic number* or *co-rank* of G. As is well known,

$$(8.43) \qquad\qquad c(G) = |E| - |V| + k(G).$$

Cf. (8.27).

8.44 Theorem [113]. *Let* $C_1, C_2, \ldots, C_c$ *be a maximal set of independent cycles in* G. *Let* $\xi_1, \xi_2, \ldots, \xi_c$ *be independent Bernoulli random variables with parameter* $\frac{1}{2}$. *Then* $\sum_i \xi_i C_i$ *is a uniform random even subgraph of* G.

Proof. Since every linear combination $\sum_i \psi_i C_i$, $\psi \in \{0, 1\}^c$, is even, and since every even graph may be expressed uniquely in this form, the uniform measure on $\{0, 1\}^c$ generates the uniform measure on $\mathcal{E}$. $\qquad\qquad\square$

One standard way of choosing such a set $C_1, C_2, \ldots, C_c$, when G is planar, is given as above by the external boundaries of the finite faces. Another is as follows. Let (V, F) be a *spanning subforest* of G, that is, the union of a spanning tree from each component of G. It is well known, and easy to check, that each edge $e_i \in E \setminus F$ can be completed by edges in F to form a unique cycle C_i. These cycles form a basis of $\mathcal{E}$. By Theorem 8.44, we may therefore find a random uniform subset of the C_j by choosing a random uniform subset of $E \setminus F$.

We show next how to couple the $q = 2$ random-cluster model and the random even subgraph of G. Let $p \in [0, \frac{1}{2}]$, and let ω be a realization of the random-cluster model on G with parameters $2p$ and $q = 2$. Let $R = (V, \gamma)$ be a uniform random even subgraph of $(V, \eta(\omega))$.

8.45 Theorem [113]. *The graph* $R = (V, \gamma)$ *is a random even subgraph of* G *with parameter* p.

This recipe for random even subgraphs provides a neat method for their simulation, provided $p \leq \frac{1}{2}$. We may sample from the random-cluster measure by the method of coupling from the past (see [203]), and then sample a uniform random even subgraph from the outcome, as above. If G is itself even, we can further sample from η_p for $p > \frac{1}{2}$ by first sampling a subgraph $(V, \widetilde{F})$ from η_{1-p} and then taking the complement $(V, E \setminus \widetilde{F})$, which has the distribution η_p. We may adapt this argument to obtain a method for sampling from η_p for $p > \frac{1}{2}$ and general G (see [113] and Exercise 8.18). When G is planar, this amounts to sampling from an antiferromagnetic Ising model on its dual graph.

There is a converse to Theorem 8.45. Take a random even subgraph (V, F) of $G = (V, E)$ with parameter $p \leq \frac{1}{2}$. To each $e \notin F$, we assign an independent random colour, blue with probability $p/(1 - p)$ and red otherwise. Let B be obtained from F by adding in all blue edges. It is left as an exercise to show that the graph (V, B) has law $\phi_{2p,2}$.

Proof of Theorem 8.45. Let $g \subseteq E$ be even, and let ω be a sample configuration of the random-cluster model on G. By the above,

$$\mathbb{P}(\gamma = g \mid \omega) = \begin{cases} 2^{-c(\omega)} & \text{if } g \subseteq \eta(\omega), \\ 0 & \text{otherwise,} \end{cases}$$

where $c(\omega) = c(V, \eta(\omega))$ is the number of independent cycles in the ω-open subgraph. Therefore,

$$\mathbb{P}(\gamma = g) = \sum_{\omega: g \subseteq \eta(\omega)} 2^{-c(\omega)} \phi_{2p,2}(\omega).$$

By (8.43),

$$\mathbb{P}(\gamma = g) \propto \sum_{\omega: g \subseteq \eta(\omega)} (2p)^{|\eta(\omega)|} (1 - 2p)^{|E \setminus \eta(\omega)|} 2^{k(\omega)} \left(\tfrac{1}{2}\right)^{|\eta(\omega)| - |V| + k(\omega)}$$

$$\propto \sum_{\omega: g \subseteq \eta(\omega)} p^{|\eta(\omega)|} (1 - 2p)^{|E \setminus \eta(\omega)|}$$

$$= [p + (1 - 2p)]^{|E \setminus g|} p^{|g|}$$

$$= p^{|g|} (1 - p)^{|E \setminus g|}, \qquad g \subseteq E.$$

The claim follows. $\qquad\qquad\qquad\qquad\qquad\qquad\qquad\qquad\qquad\qquad\qquad\quad\square$

The above account of even subgraphs would be gravely incomplete without a reminder of the so-called 'random-current representation' of the Ising model. This is a representation of the Ising measure in terms of a random field of loops and lines, and it has enabled a rigorous analysis of the Ising model. See [3, 7, 10] and [109, Chap. 9]. The random-current representation is closely related to the study of random even subgraphs.

8.7 Exercises

8.1 [119] Let $\phi_{p,q}$ be a random-cluster measure on a finite graph $G = (V, E)$ with parameters p and q. Prove that

$$\frac{d}{dp} \phi_{p,q}(A) = \frac{1}{p(1 - p)} \left\{ \phi_{p,q}(M 1_A) - \phi_{p,q}(M) \phi_{p,q}(A) \right\}$$

for any event A, where $M = |\eta(\omega)|$ is the number of open edges of a configuration ω, and 1_A is the indicator function of the event A.

8.2 (continuation) Show that $\phi_{p,q}$ is positively associated when $q \geq 1$, in that $\phi_{p,q}(A \cap B) \geq \phi_{p,q}(A)\phi_{p,q}(B)$ for increasing events A, B, but does not generally have this property when $q < 1$.

8.3 For an edge e of a graph G, we write $G \setminus e$ for the graph obtained by deleting e, and $G.e$ for the graph obtained by contracting e and identifying its endpoints. Show that the conditional random-cluster measure on G given that the edge e is closed (respectively, open) is that of $\phi_{G \setminus e, p, q}$ (respectively, $\phi_{G.e, p, q}$).

8.4 Show that random-cluster measures $\phi_{p,q}$ do not generally satisfy the BK inequality if $q > 1$. That is, find a finite graph G and increasing events A, B such that $\phi_{p,q}(A \circ B) > \phi_{p,q}(A)\phi_{p,q}(B)$.

8.5 (Important research problem, hard if true) Prove or disprove that random-cluster measures satisfy the BK inequality if $q < 1$.

8.6 Let $\phi_{p,q}$ be the random-cluster measure on a finite connected graph $G = (V, E)$. Show, in the limit as $p, q \to 0$ in such way that $q/p \to 0$, that $\phi_{p,q}$ converges weakly to the uniform spanning tree measure UST on G. Identify the corresponding limit as $p, q \to 0$ with $p = q$. Explain the relevance of these limits to the previous exercise.

8.7 [89] *Comparison inequalities.* Use the Holley inequality to prove the following 'comparison inequalities' for a random-cluster measure $\phi_{p,q}$ on a finite graph:

$$\phi_{p',q'} \leq_{\text{st}} \phi_{p,q} \qquad \text{if } q' \geq q, \ q' \geq 1, \ p' \leq p,$$

$$\phi_{p',q'} \geq_{\text{st}} \phi_{p,q} \qquad \text{if } q' \geq q, \ q' \geq 1, \ \frac{p'}{q'(1-p')} \geq \frac{p}{q(1-p)}.$$

8.8 [9] Show that the wired percolation probability $\theta^1(p, q)$ on $\mathbb{L}^d$ equals the limit of the finite-volume probabilities, in that, for $q \geq 1$,

$$\theta^1(p, q) = \lim_{\Lambda \uparrow \mathbb{Z}^d} \phi^1_{\Lambda, p, q}(0 \leftrightarrow \partial \Lambda).$$

8.9 Let $q \geq 1$ and $d \geq 3$, and consider the random-cluster measure $\psi_{L,n,p,q}$ on the slab $S(L, n) = [0, L] \times [-n, n]^{d-1}$ with free boundary conditions. Let $\Pi(p, L)$ denote the property that:

$$\liminf_{n \to \infty} \inf_{x \in S(L,n)} \{\psi_{L,n,p,q}(0 \leftrightarrow x)\} > 0.$$

Show that $\Pi(p, L) \Rightarrow \Pi(p', L')$ if $p \leq p'$ and $L \leq L'$.

8.10 [109, 180] *Mixing.* A translation τ of $\mathbb{L}^d$ induces a translation of $\Omega = \{0, 1\}^{\mathbb{E}^d}$ given by $\tau(\omega)(e) = \omega(\tau^{-1}(e))$. Let A and B be cylinder events of Ω. Show, for $q \geq 1$ and $b = 0, 1$, that

$$\phi^b_{p,q}(A \cap \tau^n B) \to \phi^b_{p,q}(A)\phi^b_{p,q}(B) \qquad \text{as } n \to \infty.$$

The following may help when $b = 0$, with a similar argument when $b = 1$.

a. Assume A is increasing. Let A be defined on the box Λ, and let Δ be a larger box with $\tau^n B$ defined on $\Delta \setminus \Lambda$. Use positive association to show that

$$\phi^0_{\Delta,p,q}(A \cap \tau^n B) \geq \phi^0_{\Lambda,p,q}(A)\phi^0_{\Delta,p,q}(\tau^n B).$$

b. Let $\Delta \uparrow \mathbb{Z}^d$, and then $n \to \infty$ and $\Lambda \uparrow \mathbb{Z}^d$, to obtain

$$\liminf_{n\to\infty} \phi^0_{p,q}(A \cap \tau^n B) \geq \phi^0_{p,q}(A)\phi^0_{p,q}(B).$$

By applying this to the complement $\overline{B}$ also, deduce that $\phi^0_{p,q}(A \cap \tau^n B) \to \phi^0_{p,q}(A)\phi^b_{p,q}(B)$.

8.11 *Ergodicity.* Deduce from the result of the previous exercise that the $\phi^b_{p,q}$ are ergodic.

8.12 Use the comparison inequalities to prove that the critical point $p_c(q)$ of the random-cluster model on $\mathbb{L}^d$ satisfies

$$p_c(1) \leq p_c(q) \leq \frac{q p_c(1)}{1 + (q-1)p_c(1)}, \qquad q \geq 1.$$

In particular, $0 < p_c(q) < 1$ if $q \geq 1$ and $d \geq 2$.

8.13 Let μ be the 'usual' coupling of the Potts measure and the random-cluster measure on a finite graph G. Derive the conditional measures of the first component given the second, and of the second given the first.

8.14 Let $q \in \{2, 3, \ldots\}$, and let $G = (V, E)$ be a finite graph. Let $W \subseteq V$, and let $\sigma_1, \sigma_2 \in \{1, 2, \ldots, q\}^W$. Starting from the random-cluster measure $\phi^W_{p,q}$ on G with members of W identified as a single point, explain how to couple the two associated Potts measures $\pi(\cdot \mid \sigma_W = \sigma_i)$, $i = 1, 2$, in such a way that: any vertex x not joined to W in the random-cluster configuration has the same spin in each of the two Potts configurations.

Let $B \subseteq \{1, 2, \ldots, q\}^Y$, where $Y \subseteq V \setminus W$. Show that

$$\left|\pi(B \mid \sigma_W = \sigma_1) - \pi(B \mid \sigma_W = \sigma_2)\right| \leq \phi^W_{p,q}(W \leftrightarrow Y).$$

8.15 *Infinite-volume coupling.* Let $\phi^b_{p,q}$ be a random-cluster measure on $\mathbb{L}^d$ with $b \in \{0, 1\}$ and $q \in \{2, 3, \ldots\}$. If $b = 0$, we assign a uniformly random element of $Q = \{1, 2, \ldots, q\}$ to each open cluster, constant within clusters and independent between. We do similarly if $b = 1$ with the difference that any infinite cluster receives spin 1. Show that the ensuing spin-measures π^b are the infinite-volume Potts measures with free and 1 boundary conditions, respectively.

8.16 *Ising mixing and ergodicity.* Using the results of the previous two exercises, or otherwise, show that the Potts measures π^b, $b = 0, 1$, are mixing (in that they satisfy the first equation of Exercise 8.10), and hence ergodic, if $\phi^b_{p,q}(0 \leftrightarrow \infty) = 0$.

8.17 [104] Show for the random-cluster model on $\mathbb{L}^2$ that $p_c(q) \geq \kappa_q$, where $\kappa_q = \sqrt{q}/(1 + \sqrt{q})$ is the self-dual point.

8.18 [113] Make a proposal for generating a random even subgraph of the graph $G = (V, E)$ with parameter p satisfying $p > \frac{1}{2}$.

You may find it useful to prove the following first. Let u, v be distinct vertices in the same component of G, and let π be a path from u to v. Let $\mathcal{F}$ be the set of even subsets of E, and $\mathcal{F}^{u,v}$ the set of subsets F such that $\deg_F(x)$ is even if and only if $x \neq u, v$. [Here, $\deg_F(x)$ is the number of elements of F incident to x.] Then $\mathcal{F}$ and $\mathcal{F}^{u,v}$ are put in one–one correspondence by $F \leftrightarrow F \bigtriangleup \pi$.

8.19 [113] Let (V, F) be a random even subgraph of $G = (V, E)$ with law η_p, where $p \leq \frac{1}{2}$. Each $e \notin F$ is coloured blue with probability $p/(1 - p)$, independently of all other edges. Let B be the union of F with the blue edges. Show that (V, B) has law $\phi_{2p,2}$.

9

Quantum Ising model

The quantum Ising model on a finite graph G may be transformed into a continuum random-cluster model on the set obtained by attaching a copy of the real line to each vertex of G. The ensuing representation of the Gibbs operator is susceptible to probabilistic analysis. One application is to an estimate of entanglement in the one-dimensional system.

9.1 The model

The quantum Ising model was introduced in [166]. Its formal definition requires a certain amount of superficially alien notation, and proceeds as follows on the finite graph $G = (V, E)$. To each vertex $x \in V$ is associated a quantum spin-$\frac{1}{2}$ with local Hilbert space $\mathbb{C}^2$. The configuration space $\mathcal{H}$ for the system is the tensor product[1] $\mathcal{H} = \bigotimes_{v \in V} \mathbb{C}^2$. As basis for the copy of $\mathbb{C}^2$ labelled by $v \in V$, we take the two eigenvectors, denoted as

$$|+\rangle_v = \begin{pmatrix} 1 \\ 0 \end{pmatrix}, \qquad |-\rangle_v = \begin{pmatrix} 0 \\ 1 \end{pmatrix},$$

of the Pauli matrix

$$\sigma_v^{(3)} = \begin{pmatrix} 1 & 0 \\ 0 & -1 \end{pmatrix}$$

at the site v, with corresponding eigenvalues ± 1. The other two Pauli matrices with respect to this basis are:

$$\sigma_v^{(1)} = \begin{pmatrix} 0 & 1 \\ 1 & 0 \end{pmatrix}, \qquad \sigma_v^{(2)} = \begin{pmatrix} 0 & -i \\ i & 0 \end{pmatrix}.$$

In the following, $|\phi\rangle$ denotes a vector and $\langle\phi|$ its adjoint (or conjugate transpose).[2]

---

[1] The tensor product $U \otimes V$ of two vector spaces over F is the dual space of the set of bilinear functionals on $U \times V$. See [99, 127].

[2] With apologies to mathematicians who dislike the bra-ket notation.

Let D be the set of $2^{|V|}$ basis vectors $|\eta\rangle$ for $\mathcal{H}$ of the form $|\eta\rangle = \bigotimes_v |\pm\rangle_v$. There is a natural one–one correspondence between D and the space

$$\Sigma = \Sigma_V = \{-1, +1\}^V.$$

We may speak of members of Σ as basis vectors, and of $\mathcal{H}$ as the Hilbert space generated by Σ.

Let $\lambda, \delta \in [0, \infty)$. The Hamiltonian of the quantum Ising model with transverse field is the matrix (or 'operator')

$$(9.1) \qquad H = -\tfrac{1}{2}\lambda \sum_{e=\langle u,v\rangle \in E} \sigma_u^{(3)}\sigma_v^{(3)} - \delta \sum_{v\in V} \sigma_v^{(1)},$$

Here, λ is the spin-coupling and δ is the transverse-field intensity. The matrix H operates on vectors (elements of $\mathcal{H}$) through the operation of each σ_v on the component of the vector at v.

Let $\beta \in [0, \infty)$ be the parameter known as 'inverse temperature'. The Hamiltonian H generates the matrix $e^{-\beta H}$, and we are concerned with the operation of this matrix on elements of $\mathcal{H}$. The right way to normalize a matrix A is by its trace

$$\mathrm{tr}(A) = \sum_{\eta\in\Sigma} \langle\eta|A|\eta\rangle.$$

Thus, we define the so-called 'density matrix' by

$$(9.2) \qquad \nu_G(\beta) = \frac{1}{Z_G(\beta)} e^{-\beta H},$$

where

$$(9.3) \qquad Z_G(\beta) = \mathrm{tr}(e^{-\beta H}).$$

It turns out that the matrix elements of $\nu_G(\beta)$ may be expressed in terms of a type of 'path integral' with respect to the continuum random-cluster model on $V \times [0, \beta]$ with parameters λ, δ, and $q = 2$. We explain this in the following two sections.

The Hamiltonian H has a unique pure *ground state* $|\psi_G\rangle$ defined at zero-temperature (that is, in the limit as $\beta \to \infty$) as the eigenvector corresponding to the lowest eigenvalue of H.

9.2 Continuum random-cluster model

The finite graph $G = (V, E)$ may be used as a base for a family of probabilistic models that live not on the vertex-set V but on the 'continuum' space $V \times \mathbb{R}$. The simplest of these models is continuum percolation, see Section

6.6. We consider here a related model called the continuum random-cluster model. Let $\beta \in (0, \infty)$, and let Λ be the 'box' $\Lambda = V \times [0, \beta]$. In the notation of Section 6.6, let $\mathbb{P}_{\Lambda,\lambda,\delta}$ denote the probability measure associated with the Poisson processes D_x, $x \in V$, and B_e, $e = \langle x, y \rangle \in E$. As sample space, we take the set Ω_Λ comprising all finite sets of cuts and bridges in Λ, and we may assume without loss of generality that no cut is the endpoint of any bridge. For $\omega \in \Omega_\Lambda$, we write $B(\omega)$ and $D(\omega)$ for the sets of bridges and cuts, respectively, of ω. The appropriate σ-field $\mathcal{F}_\Lambda$ is that generated by the open sets in the associated Skorohod topology, see [37, 83].

For a given configuration $\omega \in \Omega_\Lambda$, let $k(\omega)$ be the number of its clusters under the connection relation $\leftrightarrow$. Let $q \in (0, \infty)$, and define the 'continuum random-cluster' measure $\phi_{\Lambda,\lambda,\delta,q}$ by

$$(9.4) \qquad d\phi_{\Lambda,\lambda,\delta,q}(\omega) = \frac{1}{Z} q^{k(\omega)} d\mathbb{P}_{\Lambda,\lambda,\delta}(\omega), \qquad \omega \in \Omega_\Lambda,$$

for an appropriate normalizing constant $Z = Z_\Lambda(\lambda, \delta, q)$ called the 'partition function'. The continuum random-cluster model may be studied in much the same way as the random-cluster model on a (discrete) graph, see Chapter 8.

The space Ω_Λ is a partially ordered space with order relation given by: $\omega_1 \leq \omega_2$ if $B(\omega_1) \subseteq B(\omega_2)$ and $D(\omega_1) \supseteq D(\omega_2)$. A random variable $X : \Omega_\Lambda \to \mathbb{R}$ is called *increasing* if $X(\omega) \leq X(\omega')$ whenever $\omega \leq \omega'$. A non-empty event $A \in \mathcal{F}_\Lambda$ is called *increasing* if its indicator function 1_A is increasing. Given two probability measures μ_1, μ_2 on the measurable pair $(\Omega_\Lambda, \mathcal{F}_\Lambda)$, we write $\mu_1 \leq_{\text{st}} \mu_2$ if $\mu_1(X) \leq \mu_2(X)$ for all bounded increasing continuous random variables $X : \Omega_\Lambda \to \mathbb{R}$.

The measures $\phi_{\Lambda,\lambda,\delta,q}$ have certain properties of stochastic ordering as the parameters vary. In rough terms, the $\phi_{\Lambda,\lambda,\delta,q}$ inherit the properties of stochastic ordering and positive association enjoyed by their counterparts on discrete graphs. This will be assumed here, and the reader is referred to [40] for further details. Of value in the forthcoming Section 9.5 is the stochastic inequality

$$(9.5) \qquad \phi_{\Lambda,\lambda,\delta,q} \leq_{\text{st}} \mathbb{P}_{\Lambda,\lambda,\delta}, \qquad q \geq 1.$$

The underlying graph $G = (V, E)$ has so far been finite. Singularities emerge only in the infinite-volume (or 'thermodynamic') limit, and this may be taken in much the same manner as for the discrete random-cluster model, whenever $q \geq 1$, and for certain boundary conditions τ. Henceforth, we assume that V is a finite connected subgraph of the lattice $G = \mathbb{L}^d$, and we assign to the box $\Lambda = V \times [0, \beta]$ a suitable boundary condition. As described in [109] for the discrete case, if the boundary condition τ is chosen in such a way that the measures $\phi_{\Lambda,\lambda,\delta,q}^{\tau}$ are monotonic as $V \uparrow \mathbb{Z}^d$, then the weak limit

$$\phi_{\lambda,\delta,q,\beta}^{\tau} = \lim_{V \uparrow \mathbb{Z}^d} \phi_{\Lambda,\lambda,\delta,q}^{\tau}$$

exists. We may similarly allow the limit as $\beta \to \infty$ to obtain the 'ground state' measure

$$\phi_{\lambda,\delta,q}^{\tau} = \lim_{\beta \to \infty} \phi_{\lambda,\delta,q,\beta}^{\tau}.$$

We shall generally work with the measure $\phi_{\lambda,\delta,q}^{\tau}$ with free boundary condition τ, written simply as $\phi_{\lambda,\delta,q}$, and we note that it is sometimes appropriate to take $\beta < \infty$.

The percolation probability is given by

$$\theta(\lambda, \delta, q) = \phi_{\lambda,\delta,q}(|C| = \infty),$$

where C is the cluster at the origin $(0, 0)$, and $|C|$ denotes the aggregate (one-dimensional) Lebesgue measure of the time intervals comprising C. By re-scaling the continuum $\mathbb{R}$, we see that the percolation probability depends only on the ratio $\rho = \lambda/\delta$, and we write $\theta(\rho, q) = \theta(\lambda, \delta, q)$. The critical point is defined by

$$\rho_c(\mathbb{L}^d, q) = \sup\{\rho : \theta(\rho, q) = 0\}.$$

In the special case $d = 1$, the random-cluster model has a property of self-duality that leads to the following conjecture.

9.6 Conjecture. *The continuum random-cluster model on $\mathbb{L} \times \mathbb{R}$ with cluster-weighting factor satisfying $q \geq 1$ has critical value $\rho_c(\mathbb{L}, q) = q$.*

It may be proved by standard means that $\rho_c(\mathbb{L}, q) \geq q$. See (8.33) and [109, Sect. 6.2] for the corresponding result on the discrete lattice $\mathbb{L}^2$. The cases $q = 1, 2$ are special. The statement $\rho_c(\mathbb{L}, 1) = 1$ is part of Theorem 6.18(b). When $q = 2$, the method of so-called 'random currents' may be adapted to the quantum model with several consequences, of which we highlight the fact that $\rho_c(\mathbb{L}, 2) = 2$; see [41].

The *continuum Potts model* on $V \times \mathbb{R}$ is given as follows. Let q be an integer satisfying $q \geq 2$. To each cluster of the random-cluster model with cluster-weighting factor q is assigned a uniformly chosen 'spin' from the space $\Sigma = \{1, 2, \ldots, q\}$, different clusters receiving independent spins. The outcome is a function $\sigma : V \times \mathbb{R} \to \Sigma$, and this is the spin-vector of a 'continuum q-state Potts model' with parameters λ and δ. When $q = 2$, we refer to the model as a *continuum Ising model*.

It is not hard to see[3] that the law P of the continuous Ising model on $\Lambda = V \times [0, \beta]$ is given by

$$dP(\sigma) = \frac{1}{Z} e^{\lambda L(\sigma)} \, d\mathbb{P}_{\Lambda,\delta}(D_\sigma),$$

where D_σ is the set of $(x, s) \in V \times [0, \beta]$ such that $\sigma(x, s-) \neq \sigma(x, s+)$, $\mathbb{P}_{\Lambda,\delta}$ is the law of a family of independent Poisson processes on the time-lines $\{x\} \times [0, \beta]$, $x \in V$, with intensity δ, and

$$L(\sigma) = \sum_{\langle x,y \rangle \in E_V} \int_0^\beta 1_{\{\sigma(x,u)=\sigma(y,u)\}} \, du$$

is the aggregate Lebesgue measure of those subsets of pairs of adjacent time-lines on which the spins are equal. As usual, Z is the appropriate normalizing constant.

9.3 Quantum Ising via random-cluster

In this section, we describe the relationship between the quantum Ising model on a finite graph $G = (V, E)$ and the continuum random-cluster model on $G \times [0, \beta]$ with $q = 2$. We shall see that the density matrix $\nu_G(\beta)$ may be expressed in terms of ratios of probabilities. The basis of the following argument lies in the work of Jean Ginibre [97], and it was developed further by Campanino, von Dreyfus, Klein, and Perez. The reader is referred to [13] for more recent account. Similar geometrical transformations exist for certain other quantum models, see [14, 192].

Let $\Lambda = V \times [0, \beta]$, and let Ω_Λ be the configuration space of the continuum random-cluster model on Λ. For given λ, δ, and $q = 2$, let $\phi_{G,\beta}$ denote the corresponding continuum random-cluster measure on Ω_Λ (with free boundary conditions). Thus, for economy of notation we suppress reference to λ and δ.

We next introduce a coupling of edge and spin configurations as in Section 8.1. For $\omega \in \Omega_\Lambda$, let $S(\omega)$ denote the (finite) space of all functions $s : V \times [0, \beta] \to \{-1, +1\}$ that are constant on the clusters of ω, and let S be the union of the $S(\omega)$ over $\omega \in \Omega_\Lambda$. Given ω, we may pick an element of $S(\omega)$ uniformly at random, and we denote this random element as σ. We shall abuse notation by using $\phi_{G,\beta}$ to denote the ensuing probability measure on the coupled space $\Omega_\Lambda \times S$. For $s \in S$ and $W \subseteq V$, we write $s_{W,0}$ (respectively, $s_{W,\beta}$) for the vector $(s(x, 0) : x \in W)$ (respectively, $(s(x, \beta) : x \in W)$). We abbreviate $s_{V,0}$ and $s_{V,\beta}$ to s_0 and s_β, respectively.

---

[3] This is Exercise 9.3.

9.7 Theorem [13]. *The elements of the density matrix $\nu_G(\beta)$ satisfy*

$$(9.8) \qquad \langle \eta' | \nu_G(\beta) | \eta \rangle = \frac{\phi_{G,\beta}(\sigma_0 = \eta, \ \sigma_\beta = \eta')}{\phi_{G,\beta}(\sigma_0 = \sigma_\beta)}, \qquad \eta, \eta' \in \Sigma.$$

Readers familiar with quantum theory may recognize this as a type of Feynman–Kac representation.

Proof. We use the notation of Section 9.1. By (9.1) with $\gamma = \frac{1}{2} \sum_{\langle x,y \rangle} \lambda \mathbb{I}$ and $\mathbb{I}$ the identity matrix[4],

$$(9.9) \qquad\qquad e^{-\beta(H+\gamma)} = e^{-\beta(U+V)},$$

where

$$U = -\delta \sum_{x \in V} \sigma_x^{(1)}, \quad V = -\frac{1}{2} \sum_{e=\langle x,y \rangle \in E} \lambda(\sigma_x^{(3)} \sigma_y^{(3)} - \mathbb{I}).$$

Although these two matrices do not commute, we may use the so-called Lie–Trotter formula (see, for example, [219]) to express $e^{-\beta(U+V)}$ in terms of single-site and two-site contributions due to U and V, respectively. By the Lie–Trotter formula,

$$e^{-(U+V)\Delta t} = e^{-U\Delta t} e^{-V\Delta t} + \mathrm{O}(\Delta t^2) \qquad \text{as } \Delta t \downarrow 0,$$

so that

$$e^{-\beta(U+V)} = \lim_{\Delta t \to 0} (e^{-U\Delta t} e^{-V\Delta t})^{\beta/\Delta t}.$$

Now expand the exponential, neglecting terms of order $\mathrm{o}(\Delta t)$, to obtain

$$(9.10)$$
$$e^{-\beta(H+\gamma)} =$$

$$\lim_{\Delta t \to 0} \left(\prod_x [(1 - \delta\Delta t)\mathbb{I} + \delta\Delta t\, P_x^1] \prod_{e=\langle x,y \rangle} \left[(1 - \lambda\Delta t)\mathbb{I} + \lambda\Delta t\, P_{x,y}^3 \right] \right)^{\beta/\Delta t},$$

where $P_x^1 = \sigma_{(x)}^1 + \mathbb{I}$ and $P_{x,y}^3 = \frac{1}{2}(\sigma_x^{(3)} \sigma_y^{(3)} + \mathbb{I})$.

As noted earlier, $\Sigma = \{-1, +1\}^V$ may be considered as a basis for $\mathcal{H}$. The product (9.10) contains a collection of operators acting on sites x and on neighbouring pairs $\langle x, y \rangle$. We partition the time interval $[0, \beta]$ into N time-segments labelled $\Delta t_1, \Delta t_2, \ldots, \Delta t_N$, each of length $\Delta t \doteq \beta/N$. On neglecting terms of order $\mathrm{o}(\Delta t)$, we may see that each given time-segment arising in (9.10) contains exactly one of: the identity matrix $\mathbb{I}$, a matrix of

---

[4]Note that $\langle \eta' | e^{J+c\mathbb{I}} | \eta \rangle = e^c \langle \eta' | e^J | \eta \rangle$, so the introduction of γ into the exponent is harmless.

the form P_x^1, a matrix of the form $P_{x,y}^3$. Each such matrix occurs within the given time-segment with a certain weight.

Let us consider the actions of these matrices on the states $|\eta\rangle$ for each time interval Δt_i, $i \in \{1, 2, \ldots, N\}$. The matrix elements of the single-site operator at x are given by

$$(9.11) \qquad \langle \eta' | \sigma_x^{(1)} + \mathbb{I} | \eta \rangle \equiv 1.$$

This is easily checked by exhaustion. When this matrix occurs in some time-segment Δt_i, we place a mark in the interval $\{x\} \times \Delta t_i$, and we call this mark a *cut*. Such a cut has a corresponding weight $\delta \Delta t + o(\Delta t)$.

The matrix element involving the neighbouring pair $\langle x, y \rangle$ yields, as above,

$$(9.12) \qquad \tfrac{1}{2} \langle \eta' | \sigma_x^{(3)} \sigma_y^{(3)} + \mathbb{I} | \eta \rangle = \begin{cases} 1 & \text{if } \eta_x = \eta_y = \eta_x' = \eta_y', \\ 0 & \text{otherwise.} \end{cases}$$

When this occurs in some time-segment Δt_i, we place a *bridge* between the intervals $\{x\} \times \Delta t_i$ and $\{y\} \times \Delta t_i$. Such a bridge has a corresponding weight $\lambda \Delta t + o(\Delta t)$.

In the limit $\Delta t \to 0$, the spin operators generate thus a Poisson process with intensity δ of cuts in each time-line $\{x\} \times [0, \beta]$, and a Poisson process with intensity λ of bridges between each pair $\{x\} \times [0, \beta]$, $\{y\} \times [0, \beta]$ of time-lines, for neighbouring x and y. These Poisson processes are independent of one another. We write D_x for the set of cuts at the site x, and B_e for the set of bridges corresponding to an edge $e = \langle x, y \rangle$. The configuration space is the set Ω_Λ containing all finite sets of cuts and bridges, and we may assume without loss of generality that no cut is the endpoint of any bridge.

For two points $(x, s), (y, t) \in \Lambda$, we write as before $(x, s) \leftrightarrow (y, t)$ if there exists a cut-free path from the first to the second that traverses time-lines and bridges. A *cluster* is a maximal subset C of Λ such that $(x, s) \leftrightarrow (y, t)$ for all $(x, s), (y, t) \in C$. Thus the connection relation $\leftrightarrow$ generates a continuum percolation process on Λ, and we write $\mathbb{P}_{\Lambda, \lambda, \delta}$ for the probability measure corresponding to the weight function on the configuration space Ω_Λ. That is, $\mathbb{P}_{\Lambda, \lambda, \delta}$ is the measure governing a family of independent Poisson processes of cuts (with intensity δ) and of bridges (with intensity λ). The ensuing percolation process has appeared in Section 6.6.

Equations (9.11)–(9.12) are to be interpreted in the following way. In calculating the operator $e^{-\beta(H+\gamma)}$, we average over contributions from realizations of the Poisson processes, on the basis that the quantum spins are constant on every cluster of the corresponding percolation process, and each such spin-function is equiprobable.

More explicitly,

(9.13)
$$e^{-\beta(H+\gamma)} = \int d\mathbb{P}_{\Lambda,\lambda,\delta}(\omega)\left(\mathcal{T} \prod_{(x,t)\in D} P_x^1(t) \prod_{(\langle x,y\rangle,t')\in B} P_{x,y}^3(t')\right),$$

where $\mathcal{T}$ denotes the time-ordering of the terms in the products, and B (respectively, D) is the set of all bridges (respectively, cuts) of the configuration $\omega \in \Omega_\Lambda$.

Let $\omega \in \Omega_\Lambda$. Let μ_ω be the counting measure on the space $S(\omega)$ of functions $s : V \times [0, \beta] \to \{-1, +1\}$ that are constant on the clusters of ω. Let $K(\omega)$ be the time-ordered product of operators in (9.13). We may evaluate the matrix elements of $K(\omega)$ by inserting the 'resolution of the identity'

(9.14)
$$\sum_{\eta\in\Sigma} |\eta\rangle\langle\eta| = \mathbb{I}$$

between any two factors in the product, obtaining by (9.11)–(9.12) that

(9.15) $$\langle\eta'|K(\omega)|\eta\rangle = \sum_{s\in S(\omega)} 1_{\{s_0=\eta\}} 1_{\{s_\beta=\eta'\}}, \qquad \eta, \eta' \in \Sigma.$$

This is the number of spin-allocations to the clusters of ω with given spin-vectors at times 0 and β.

The matrix elements of $\nu_G(\beta)$ are therefore given by

(9.16) $$\langle\eta'|\nu_G(\beta)|\eta\rangle = \frac{1}{Z_{G,\beta}} \int 1_{\{s_0=\eta\}} 1_{\{s_\beta=\eta'\}} \, d\mu_\omega(s) \, d\mathbb{P}_{\Lambda,\lambda,\delta}(\omega),$$

for $\eta, \eta' \in \Sigma$, where

(9.17) $$Z_{G,\beta} = \mathrm{tr}(e^{-\beta(H+\gamma)}).$$

For $\eta, \eta' \in \Sigma$, let $I_{\eta,\eta'}$ be the indicator function of the event (in Ω_Λ) that, for all $x, y \in V$,

$$\text{if } (x, 0) \leftrightarrow (y, 0), \text{ then } \eta_x = \eta_y,$$
$$\text{if } (x, \beta) \leftrightarrow (y, \beta), \text{ then } \eta'_x = \eta'_y,$$
$$\text{if } (x, 0) \leftrightarrow (y, \beta), \text{ then } \eta_x = \eta'_y.$$

This is the event that the pair (η, η') of initial and final spin-vectors is

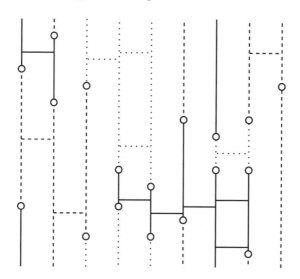

Figure 9.1. An example of a space–time configuration contributing to the Poisson integral (9.18). The cuts are shown as circles and the distinct connected clusters are indicated with different line-types.

'compatible' with the random-cluster configuration. We have that

$$
(9.18) \quad \langle \eta' | \nu_G(\beta) | \eta \rangle = \frac{1}{Z_{G,\beta}} \int d\mathbb{P}_{\Lambda,\lambda,\delta}(\omega) \sum_{s \in S(\omega)} 1_{\{s_0 = \eta\}} 1_{\{s_\beta = \eta'\}}
$$

$$
= \frac{1}{Z_{G,\beta}} \int 2^{\bar{k}(\omega)} I_{\eta,\eta'} \, d\mathbb{P}_{\Lambda,\lambda,\delta}(\omega)
$$

$$
= \frac{1}{Z_{G,\beta}} \phi_{G,\beta}(\sigma_0 = \eta, \ \sigma_\beta = \eta'). \qquad \eta, \eta' \in \Sigma,
$$

where $\bar{k}(\omega)$ is the number of clusters of ω containing no point of the form $(v, 0)$ or (v, β), for $v \in V$. See Figure 9.1 for an illustration of the space–time configurations contributing to the Poisson integral (9.18).

On setting $\eta = \eta'$ in (9.18) and summing over $\eta \in \Sigma$, we find that

$$
(9.19) \qquad\qquad Z_{G,\beta} = \phi_{G,\beta}(\sigma_0 = \sigma_\beta),
$$

as required. □

This section closes with an alternative expression for the trace formula for $Z_{G,\beta} = \mathrm{tr}(e^{-\beta(H+\gamma)})$. We consider 'periodic' boundary conditions on Λ obtained by, for each $x \in V$, identifying the pair $(x, 0)$ and (x, β) of points. Let $k^{\mathrm{per}}(\omega)$ be the number of open clusters of ω with periodic boundary conditions, and $\phi_{G,\beta}^{\text{per}}$ be the corresponding random-cluster measure. By setting $\eta' = \eta$ in (9.18) and summing,

$$(9.20) \quad 1 = \sum_{\eta \in \Sigma} \langle \eta | \nu_G(\beta) | \eta \rangle = \frac{1}{Z_{G,\beta}} \int 2^{\bar{k}(\omega)} 2^{k^{\text{per}}(\omega) - \bar{k}(\omega)} \, d\mathbb{P}_{\Lambda,\lambda,\delta}(\omega),$$

whence $Z_{G,\beta}$ equals the normalizing constant for the periodic random-cluster measure $\phi_{G,\beta}^{\text{per}}$.

9.4 Long-range order

The density matrix has been expressed in terms of the continuous random-cluster model. This representation incorporates a relationship between the phase transitions of the two models. The so-called 'order parameter' of the random-cluster model is of course its percolation probability θ, and the phase transition takes place at the point of singularity of θ. Another way of expressing this is to say that the two-point connectivity function

$$\tau_{G,\beta}(x, y) = \phi_{G,\beta}^{\text{per}}\big((x, 0) \leftrightarrow (y, 0)\big), \qquad x, y \in V,$$

is a natural measure of long-range order in the random-cluster model. It is less clear how best to summarize the concept of long-range order in the quantum Ising model, and, for reasons that are about to become clear, we use the quantity

$$\text{tr}\big(\nu_G(\beta) \sigma_x^{(3)} \sigma_y^{(3)}\big), \qquad x, y \in V.$$

9.21 Theorem [13]. *Let $G = (V, E)$ be a finite graph, and $\beta > 0$. We have that*

$$\tau_{G,\beta}(x, y) = \text{tr}\big(\nu_G(\beta) \sigma_x^{(3)} \sigma_y^{(3)}\big), \qquad x, y \in V.$$

Proof. The argument leading to (9.18) is easily adapted to obtain

$$\text{tr}\big(\nu_G(\beta) \cdot \tfrac{1}{2}(\sigma_x^{(3)} \sigma_y^{(3)} + \mathbb{I})\big) = \frac{1}{Z_{G,\beta}} \int 2^{\bar{k}(\omega)} \bigg(\sum_{\eta:\, \eta_x = \eta_y} I_{\eta,\eta} \bigg) d\mathbb{P}_{\Lambda,\lambda,\delta}(\omega).$$

Now,

$$\sum_{\eta:\, \eta_x = \eta_y} I_{\eta,\eta} = \begin{cases} 2^{k^{\text{per}}(\omega) - \bar{k}(\omega)} & \text{if } (x, 0) \leftrightarrow (y, 0), \\ 2^{k^{\text{per}}(\omega) - \bar{k}(\omega) - 1} & \text{if } (x, 0) \nleftrightarrow (y, 0), \end{cases}$$

whence, by the remark at the end of the last section,

$$\text{tr}\big(\nu_G(\beta) \cdot \tfrac{1}{2}(\sigma_x^{(3)} \sigma_y^{(3)} + \mathbb{I})\big) = \tau_{G,\beta}(x, y) + \tfrac{1}{2}(1 - \tau_{G,\beta}(x, y)),$$

and the claim follows. $\square$

The infinite-volume limits of the quantum Ising model on G are obtained in the 'ground state' as $\beta \to \infty$, and in the spatial limit as $|V| \to \infty$. The paraphernalia of the discrete random-cluster model may be adapted to the current continuous setting in order to understand the issues of existence and uniqueness of these limits. This is not investigated here. Instead, we point out that the behaviour of the two-point connectivity function, after taking the limits $\beta \to \infty$, $|V| \to \infty$, depends pivotally on the existence or not of an unbounded cluster in the infinite-volume random-cluster model. Let $\phi_{\lambda,\delta,2}$ be the infinite-volume measure, and let

$$\theta(\lambda, \delta) = \phi_{\lambda,\delta,2}(C_0 \text{ is unbounded})$$

be the percolation probability. Then $\tau_{\lambda,\delta}(x, y) \to 0$ as $|x - y| \to \infty$, when $\theta(\lambda, \delta) = 0$. On the other hand, by the FKG inequality and the (a.s.) uniqueness of the unbounded cluster,

$$\tau_{\lambda,\delta}(x, y) \geq \theta(\lambda, \delta)^2,$$

implying that $\tau_{\lambda,\delta}(x, y)$ is bounded uniformly away from 0 when $\theta(\lambda, \delta) > 0$. Thus the critical point of the random-cluster model is also a point of phase transition for the quantum model.

A more detailed investigation of the infinite-volume limits and their implications for the quantum Ising model may be found in [13]. As pointed out there, the situation is more interesting in the 'disordered' setting, when the λ_e and δ_x are themselves random variables.

A principal technique for the study of the classical Ising model is the so-called random-current method. This may be adapted to a 'random-parity representation' for the continuum Ising model corresponding to the continuous random-cluster model of Section 9.3, see [41, 69]. Many results follow for the quantum Ising model in a general number of dimensions, see [41].

9.5 Entanglement in one dimension

It is shown next how the random-cluster analysis of the last section enables progress with the problem of so-called 'quantum entanglement' in one dimension. The principle reference for the work of this section is [118].

Let $G = (V, E)$ be a finite graph, and let $W \subseteq V$. A considerable effort has been spent on understanding the so-called 'entanglement' of the spins in W relative to those of $V \setminus W$, in the (ground state) limit as $\beta \to \infty$. This is already a hard problem when G is a finite subgraph of the line $\mathbb{L}$. Various methods have been used in this case, and a variety of results, some rigorous, obtained.

The first step in the definition of entanglement is to define the *reduced density matrix*

$$v_G^W(\beta) = \text{tr}_{V\backslash W}(v_G(\beta)),$$

where the trace is taken over the Hilbert space $\mathcal{H}_{V\backslash W} = \bigotimes_{x \in V\backslash W} \mathbb{C}^2$ of spins of vertices of $V \backslash W$. An analysis (omitted here) exactly parallel to that leading to Theorem 9.7 allows the following representation of the matrix elements of $v_G^W(\beta)$.

9.22 Theorem [118]. *The elements of the reduced density matrix $v_G^W(\beta)$ satisfy*
(9.23)
$$\langle \eta' | v_G^W(\beta) | \eta \rangle = \frac{\phi_{G,\beta}(\sigma_{W,0} = \eta,\ \sigma_{W,\beta} = \eta' \mid F)}{\phi_{G,\beta}(\sigma_0 = \sigma_\beta \mid F)}, \qquad \eta, \eta' \in \Sigma_W,$$

where F is the event that $\sigma_{V\backslash W,0} = \sigma_{V\backslash W,\beta}$.

Let D_W be the set of $2^{|W|}$ vectors $|\eta\rangle$ of the form $|\eta\rangle = \bigotimes_{w \in W} |\pm\rangle_w$, and write $\mathcal{H}_W$ for the Hilbert space generated by D_W. Just as before, there is a natural one–one correspondence between D_W and the space $\Sigma_W = \{-1, +1\}^W$, and we shall thus regard $\mathcal{H}_W$ as the Hilbert space generated by Σ_W.

We may write

$$v_G = \lim_{\beta\to\infty} v_G(\beta) = |\psi_G\rangle\langle\psi_G|$$

for the density matrix corresponding to the ground state of the system, and similarly

(9.24) $$v_G^W = \text{tr}_{V\backslash W}(|\psi_G\rangle\langle\psi_G|) = \lim_{\beta\to\infty} v_G^W(\beta).$$

The entanglement of the spins in W may be defined as follows.

9.25 Definition. The *entanglement* of the spins of W relative to its complement $V \backslash W$ is the entropy

(9.26) $$S_G^W = -\text{tr}(v_G^W \log_2 v_G^W).$$

The behaviour of S_G^W, for general G and W, is not understood at present. We specialize here to the case of a finite subset of the one-dimensional lattice $\mathbb{L}$. Let $m, L \geq 0$ and take $V = [-m, m+L]$ and $W = [0, L]$, viewed as subsets of $\mathbb{Z}$. We obtain the graph G from V by adding edges between each pair $x, y \in V$ with $|x - y| = 1$. We write $v_m(\beta)$ for $v_G(\beta)$, and S_m^L (respectively, v_m^L) for S_G^W (respectively, v_G^W). A key step in the study of S_m^L

for large m is a bound on the norm of the difference $v_m^L - v_n^L$. The *operator norm* of a Hermitian matrix[5] A is given by

$$\|A\| = \sup_{\|\psi\|=1} |\langle \psi | A | \psi \rangle|,$$

where the supremum is over all vectors ψ with L^2-norm 1.

9.27 Theorem [40, 118]. *Let* $\lambda, \delta \in (0, \infty)$ *and write* $\rho = \lambda/\delta$. *There exist constants* C, α, γ *depending on* ρ *and satisfying* $\gamma > 0$ *when* $\rho < 2$ *such that*

$$(9.28) \quad \|v_m^L - v_n^L\| \le \min\{2, CL^\alpha e^{-\gamma m}\}, \qquad 2 \le m \le n < \infty, \ L \ge 1.$$

This was proved in [118] for $\rho < 1$, and the stronger result follows from the identification of the critical point $\rho_c = 2$ of [41]. The constant γ is, apart from a constant factor, the reciprocal of the correlation length of the associated random-cluster model.

Inequality (9.28) is proved by the following route. Consider the continuum random-cluster model with $q = 2$ on the space–time graph $\Lambda = V \times [0, \beta]$ with 'partial periodic top/bottom boundary conditions'; that is, for each $x \in V \setminus W$, we identify the two points $(x, 0)$ and (x, β). Let $\phi_{m,\beta}^{\mathrm{p}}$ denote the associated random-cluster measure on Ω_Λ. To each cluster of $\omega \in \Omega_\Lambda$ we assign a random spin from $\{-1, +1\}$ in the usual manner, and we abuse notation by using $\phi_{m,\beta}^{\mathrm{p}}$ to denote the measure governing both the random-cluster configuration and the spin configuration. Let

$$a_{m,\beta} = \phi_{m,\beta}^{\mathrm{p}}(\sigma_{W,0} = \sigma_{W,\beta}),$$

noting that

$$a_{m,\beta} = \phi_{m,\beta}(\sigma_0 = \sigma_\beta \mid F)$$

as in (9.23).

By Theorem 9.22,

$$(9.29) \quad \langle \psi | v_m^L(\beta) - v_n^L(\beta) | \psi \rangle$$
$$= \frac{\phi_{m,\beta}^{\mathrm{p}}(c(\sigma_{W,0})\overline{c(\sigma_{W,\beta})})}{a_{m,\beta}} - \frac{\phi_{n,\beta}^{\mathrm{p}}(c(\sigma_{W,0})\overline{c(\sigma_{W,\beta})})}{a_{n,\beta}},$$

where $c : \{-1, +1\}^W \to \mathbb{C}$ and

$$\psi = \sum_{\eta \in \Sigma_W} c(\eta)\eta \in \mathcal{H}_W.$$

---

[5]A matrix is called *Hermitian* if it equals its conjugate transpose.

The property of ratio weak-mixing (for a random-cluster measure ϕ) is used in the derivation of (9.28) from (9.29). This may be stated roughly as follows. Let A and B be events in the continuum random-cluster model that are defined on regions R_A and R_B of space, respectively. What can be said about the difference $\phi(A \cap B) - \phi(A)\phi(B)$ when the distance $d(R_A, R_B)$ between R_A and R_B is large? It is not hard to show that this difference is exponentially small in the distance, so long as the random-cluster model has exponentially decaying connectivities, and such a property is called 'weak mixing'. It is harder to show a similar bound for the difference $\phi(A \mid B) - \phi(A)$, and such a bound is termed 'ratio weak-mixing'. The ratio weak-mixing property of random-cluster measures has been investigated in [19, 20] for the discrete case and in [118] for the continuum model.

At the final step of the proof of Theorem 9.27, the random-cluster model is compared via (9.5) with the continuum percolation model of Section 6.6, and the exponential decay of Theorem 9.27 follows by Theorem 6.18. A logarithmic bound on the entanglement entropy follows for sufficiently small λ/δ.

9.30 Theorem [118]. *Let* $\lambda, \delta \in (0, \infty)$ *and write* $\rho = \lambda/\delta$. *There exists* $\rho_0 \in (0, 2]$ *such that: for* $\rho < \rho_0$, *there exists* $K = K(\rho) < \infty$ *such that*

$$S_m^L \leq K \log_2 L, \qquad m \geq 0, \ L \geq 2.$$

Here is the idea of the proof. Theorem 9.27 implies, by a classic theorem of Weyl, that the spectra (and hence the entropies) of ν_m^L and ν_n^L are close to one another. It is an easy calculation that $S_m^L \leq c \log L$ for $m \leq c' \log L$, and the conclusion follows.

A stronger result is known to physicists, namely that the entanglement S_m^L is bounded above, uniformly in L, whenever ρ is sufficiently small, and perhaps for all $\rho < \rho_c$, where $\rho_c = 2$ is the critical point. It is not clear whether this is provable by the methods of this chapter. See Conjecture 9.6 above, and the references in [118].

There is no rigorous picture known of the behaviour of S_m^L for large ρ, or of the corresponding quantity in dimensions $d \geq 2$, although Theorem 9.27 has a counterpart in these settings. Theorem 9.30 may be extended to the *disordered* system in which the intensities λ, δ are independent random variables indexed by the vertices and edges of the underlying graph, subject to certain conditions on these variables (cf. Theorem 6.19 and the preceding discussion).

9.6 Exercises

9.1 Explain in what manner the continuum random-cluster measure $\phi_{\lambda,\delta,q}$ on $\mathbb{L} \times \mathbb{R}$ is 'self-dual' when $\rho = \lambda/\delta$ satisfies $\rho = q$.

9.2 (continuation) Show that the critical value of ρ satisfies $\rho_c \geq q$ when $q \geq 1$.

9.3 Let $\phi_{\lambda,\delta,q}$ be the continuum random-cluster measure on $G \times [0, \beta]$, where G is a finite graph, $\beta < \infty$, and $q \in \{2, 3, \dots\}$. To each cluster is assigned a spin chosen uniformly at random from the set $\{1, 2, \dots, q\}$, these spins being constant within clusters and independent between them. Find an expression for the law of the ensuing (Potts) spin-process on $V \times [0, \beta]$.

10

Interacting particle systems

The contact, voter, and exclusion models are Markov processes in continuous time with state space $\{0, 1\}^V$ for some countable set V. In the *voter model*, each element of V may be in either of two states, and its state flips at a rate that is a weighted average of the states of the other elements. Its analysis hinges on the recurrence or transience of an associated Markov chain. When $V = \mathbb{Z}^2$ and the model is generated by simple random walk, the only invariant measures are the two point masses on the (two) states representing unanimity. The picture is more complicated when $d \geq 3$. In the *exclusion model*, a set of particles moves about V according to a 'symmetric' Markov chain, subject to exclusion. When $V = \mathbb{Z}^d$ and the Markov chain is translation-invariant, the product measures are invariant for this process, and furthermore these are exactly the extremal invariant measures. The chapter closes with a brief account of the *stochastic Ising model*.

10.1 Introductory remarks

There are many beautiful problems of physical type that may be modelled as Markov processes on the compact state space $\Sigma = \{0, 1\}^V$ for some countable set V. Amongst the most studied to date by probabilists are the contact, voter, and exclusion models, and the stochastic Ising model. This significant branch of modern probability theory had its nascence around 1970 in the work of Roland Dobrushin, Frank Spitzer, and others, and has been brought to maturity through the work of Thomas Liggett and colleagues. The basic references are Liggett's two volumes [167, 169], see also [170].

The general theory of Markov processes, with its intrinsic complexities, is avoided here. The first three processes of this chapter may be constructed via 'graphical representations' involving independent random walks. There is a general approach to such important matters as the existence of processes, for an account of which the reader is referred to [167]. The two observations of note are that the state space Σ is compact, and that the Markov processes

$(\eta_t : t \geq 0)$ of this section are Feller processes, which is to say that the transition measures are weakly continuous functions of the initial state.[1]

For a given Markov process, the two main questions are to identify the set of invariant measures, and to identify the 'basin of attraction' of a given invariant measure. The processes of this chapter will possess a non-empty set I of invariant measures, although it is not always possible to describe all members of this set explicitly. Since I is a convex set of measures, it suffices to describe its extremal elements. We shall see that, in certain circumstances, $|I| = 1$, and this may be interpreted as the absence of long-range order.

Since V is infinite, Σ is uncountable. We normally specify the transition operators of a Markov chain on such Σ by specifying its *generator*. This is an operator $\mathcal{L}$ acting on an appropriate dense subset of $C(\Sigma)$, the space of continuous functions on Σ endowed with the product topology and the supremum norm. It is determined by its values on the space $C(\Sigma)$ of cylinder functions, being the set of functions that depend on only finitely many coordinates in Σ. For $f \in C(\Sigma)$, we write $\mathcal{L}f$ in the form

$$(10.1) \qquad \mathcal{L}f(\eta) = \sum_{\eta' \in \Sigma} c(\eta, \eta')[f(\eta') - f(\eta)], \qquad \eta \in \Sigma,$$

for some function c sometimes called the 'speed (or rate) function'. For $\eta \neq \eta'$, we think of $c(\eta, \eta')$ as being the rate at which the process, when in state η, jumps to state η'.

The processes η_t possesses a transition semigroup $(S_t : t \geq 0)$ acting on $C(\Sigma)$ and given by

$$(10.2) \qquad S_t f(\eta) = \mathbb{E}^{\eta}(f(\eta_t)), \qquad \eta \in \Sigma,$$

where $\mathbb{E}^{\eta}$ denotes expectation under the assumption $\eta_0 = \eta$. Under certain conditions on the process, the transition semigroup is related to the generator by the formula

$$(10.3) \qquad S_t = \exp(t\mathcal{L}),$$

suitably interpreted according to the Hille–Yosida theorem, see [167, Sect. I.2]. The semigroup acts on probability measures by

$$(10.4) \qquad \mu S_t(A) = \int_{\Sigma} \mathbb{P}^{\eta}(\eta_t \in A) \, d\mu(\eta).$$

---

[1] Let $C(\Sigma)$ denote the space of continuous functions on Σ endowed with the product topology and the supremum norm. The process η_t is called *Feller* if, for $f \in C(\Sigma)$, $f_t(\eta) = \mathbb{E}^{\eta}(f(\eta_t))$ defines a function belonging to $C(\Sigma)$. Here, $\mathbb{E}^{\eta}$ denotes expectation with initial state η.

A probability measure μ on Σ is called *invariant* for the process η_t if $\mu S_t = \mu$ for all t. Under suitable conditions, μ is invariant if and only if

$$(10.5) \qquad \int \mathcal{L}f \, d\mu = 0 \qquad \text{for all } f \in \mathcal{C}(\Sigma).$$

In the remainder of this chapter, we shall encounter certain constructions of Markov processes on Σ, and all such constructions will satisfy the conditions alluded to above.

10.2 Contact process

Let $G = (V, E)$ be a connected graph with bounded vertex-degrees. The state space is $\Sigma = \{0, 1\}^V$, where the local state 1 (respectively, 0) represents 'ill' (respectively, 'healthy'). Ill vertices recover at rate δ, and healthy vertices become ill at a rate that is linear in the number of ill neighbours. See Chapter 6.

We proceed more formally as follows. For $\eta \in \Sigma$ and $x \in V$, let η_x denote the state obtained from η by flipping the local state of x. That is,

$$(10.6) \qquad \eta_x(y) = \begin{cases} 1 - \eta(x) & \text{if } y = x, \\ \eta(y) & \text{otherwise.} \end{cases}$$

We let the function c of (10.1) be given by

$$c(\eta, \eta_x) = \begin{cases} \delta & \text{if } \eta(x) = 1, \\ \lambda|\{y \sim x : \eta(y) = 1\}| & \text{if } \eta(x) = 0, \end{cases}$$

where λ and δ are strictly positive constants. If $\eta' = \eta_x$ for no $x \in V$, and $\eta' \neq \eta$, we set $c(\eta, \eta') = 0$.

We saw in Chapter 6 that the point mass on the empty set, $\underline{v} = \delta_\varnothing$, is the minimal invariant measure of the process, and that there exists a maximal invariant measure $\overline{v}$ obtained as the weak limit of the process with initial state V. As remarked at the end of Section 6.3, when $G = \mathbb{L}^d$, the set of extremal invariant measures is exactly $\mathcal{I}_e = \{\delta_\varnothing, \overline{v}\}$, and $\delta_\varnothing = \overline{v}$ if and only if there is no percolation in the associated oriented percolation model in continuous time. Of especial use in proving these facts was the coupling of contact models in terms of Poisson processes of cuts and (directed) bridges.

We revisit duality briefly, see Theorem 6.1. For $\eta \in \Sigma$ and $A \subseteq V$, let

$$(10.7) \quad H(\eta, A) = \prod_{x \in A}[1 - \eta(x)] = \begin{cases} 1 & \text{if } \eta(x) = 0 \text{ for all } x \in A, \\ 0 & \text{otherwise.} \end{cases}$$

The conclusion of Theorem 6.1 may be expressed more generally as

$$\mathbb{E}^A(H(A_t, B)) = \mathbb{E}^B(H(A, B_t)),$$

where A_t (respectively, B_t) denotes the contact model with initial state $A_0 = A$ (respectively, $B_0 = B$). This may seem a strange way to express the duality relation, but its significance may become clearer soon.

10.3 Voter model

Let V be a countable set, and let $P = (p_{x,y} : x, y \in V)$ be the transition matrix of a Markov chain on V. The associated voter model is given by choosing

$$(10.8) \qquad c(\eta, \eta_x) = \sum_{y: \eta(y) \neq \eta(x)} p_{x,y}$$

in (10.1). The meaning of this is as follows. Each member of V is an individual in a population, and may have either of two opinions at any given time. Let $x \in V$. At times of a rate-1 Poisson process, x selects a random y according to the measure $p_{x,y}$, and adopts the opinion of y. It turns out that the behaviour of this model is closely related to the transience/recurrence of the chain with transition matrix matrix P, and of properties of its harmonic functions.

The voter model has two absorbing states, namely all 0 and all 1, and we denote by δ_0 and δ_1 the point masses on these states. Any convex combination of δ_0 and δ_1 is invariant also, and thus we ask for conditions under which every invariant measure is of this form. A duality relation will enable us to answer this question.

It is helpful to draw the graphical representation of the process. With each $x \in V$ is associated a 'time-line' $[0, \infty)$, and on each such time-line is marked the set of epochs of a Poisson process Po_x with intensity 1. Different time-lines possess independent Poisson processes. Associated with each epoch of the Poisson process at x is a vertex y chosen at random according to the transition matrix P. The choice of y has the interpretation given above.

Consider the state of vertex x at time t. We imagine a particle that is at position x at time t, and we write $X_x(0) = x$. When we follow the time-line $x \times [0, t]$ backwards in time, that is, from the point (x, t) towards the point $(x, 0)$, we encounter a first point (first in this reversed ordering of time) belonging to Po_x. At this time, the particle jumps to the selected neighbour of x. Continuing likewise, the particle performs a simple random walk about V. Writing $X_x(t)$ for its position at time 0, the (voter) state of x at time t is precisely that of $X_x(t)$ at time 0.

Suppose we proceed likewise starting from two vertices x and y at time t. Tracing the states of x and y backwards, each follows a Markov chain with transition matrix P, denoted X_x and X_y respectively. These chains are independent until the first time (if ever) they meet. When they meet, they 'coalesce': if they ever occupy the same vertex at any given time, then they follow the same trajectory subsequently.

We state this as follows. The presentation here is somewhat informal, and may be made more complete as in [167]. We write $(\eta_t : t \geq 0)$ for the voter process, and $\mathcal{S}$ for the set of finite subsets of V.

10.9 Theorem. *Let $A \in \mathcal{S}$, $\eta \in \Sigma$, and let $(A_t : t \geq 0)$ be a system of coalescing random walks beginning on the set $A_0 = A$. Then,*

$$\mathbb{P}^\eta(\eta_t \equiv 1 \text{ on } A) = \mathbb{P}^A(\eta \equiv 1 \text{ on } A_t), \qquad t \geq 0.$$

This may be expressed in the form

$$\mathbb{E}^\eta(H(\eta_t, A)) = \mathbb{E}^A(H(\eta, A_t)),$$

with

$$H(\eta, A) = \prod_{x \in A} \eta(x).$$

Proof. Each side of the equation is the measure of the complement of the event that, in the graphical representation, there is a path from $(x, 0)$ to (a, t) for some x with $\eta(x) = 0$ and some $a \in A$.   □

For simplicity, we restrict ourselves henceforth to a case of special interest, namely with V the vertex-set $\mathbb{Z}^d$ of the d-dimensional lattice $\mathbb{L}^d$ with $d \geq 1$, and with $p_{x,y} = p(x - y)$ for some function p. In the special case of simple random walk, where

(10.10) $$p(z) = \frac{1}{2d}, \qquad z \text{ a neighbour of } 0,$$

we have that $\eta(x)$ flips at a rate equal to the proportion of neighbours of x whose states disagree with the current value $\eta(x)$. The case of general P is treated in [167].

Let X_t and Y_t be independent random walks on $\mathbb{Z}^d$ with rate-1 exponential holding times, and jump distribution $p_{x,y} = p(y - x)$. The difference $X_t - Y_t$ is a Markov chain also. If $X_t - Y_t$ is recurrent, we say that we are in the *recurrent case*, otherwise the *transient case*. The analysis of the voter model is fairly simple in the recurrent case.

10.11 Theorem. *Assume we are in the recurrent case.*

(a) $\mathcal{I}_e = \{\delta_0, \delta_1\}$.

(b) *If μ is a probability measure on Σ with $\mu(\eta(x) = 1) = \alpha$ for all $x \in \mathbb{Z}^d$, then $\mu S_t \Rightarrow (1 - \alpha)\delta_0 + \alpha\delta_1$ as $t \to \infty$.*

The situation is quite different in the transient case. We may construct a family of distinct invariant measures ν_α indexed by $\alpha \in [0, 1]$, and we do this as follows. Let ϕ_α be product measure on Σ with density α. We shall show the existence of the weak limits $\nu_\alpha = \lim_{t\to\infty} \phi_\alpha S_t$, and it turns out that the ν_α are exactly the extremal invariant measures. A partial proof of the next theorem is provided below.

10.12 Theorem. *Assume we are in the transient case.*

(a) *The weak limits $\nu_\alpha = \lim_{t\to\infty} \phi_\alpha S_t$ exist.*

(b) *The ν_α are translation-invariant and ergodic[2], with density*

$$\nu_\alpha(\eta(x) = 1) = \alpha, \qquad x \in \mathbb{Z}^d.$$

(c) $\mathcal{I}_e = \{\nu_\alpha : \alpha \in [0, 1]\}$.

We return briefly to the voter model corresponding to simple random walk on $\mathbb{L}^d$, see (10.10). It is an elementary consequence of Pólya's theorem, Theorem 1.32, that we are in the recurrent case if and only $d \leq 2$.

Proof of Theorem 10.11. By assumption, we are in the recurrent case. Let $x, y \in \mathbb{Z}^d$. By duality and recurrence,

$$(10.13) \quad \mathbb{P}(\eta_t(x) \neq \eta_t(y)) \leq \mathbb{P}\big(X_x(u) \neq X_y(u) \text{ for } 0 \leq u \leq t\big)$$
$$\to 0 \qquad \text{as } t \to \infty.$$

For $A \in \mathcal{S}, A \neq \varnothing$,

$$\mathbb{P}(\eta_t \text{ is non-constant on } A) \leq \mathbb{P}^A(|A_t| > 1),$$

and, by (10.13),

$$(10.14) \quad \mathbb{P}^A(|A_t| > 1) \leq \sum_{x, y \in A} \mathbb{P}\big(X_x(u) \neq X_y(u) \text{ for } 0 \leq u \leq t\big)$$
$$\to 0 \qquad \text{as } t \to \infty.$$

It follows that, for any invariant measure μ, the μ-measure of the set of constant configurations is 1. Only the convex combinations of δ_0 and δ_1 have this property.

---

[2] A probability measure μ on Σ is *ergodic* if any shift-invariant event has μ-probability either 0 or 1. It is standard that the ergodic measures are extremal within the class of translation-invariant measures, see [94] for example.

Let μ be a probability measure with density α, as in the statement of the theorem, and let $A \in \mathcal{S}$, $A \neq \varnothing$. By Theorem 10.9,

$$\mu S_t(\{\eta : \eta \equiv 1 \text{ on } A\}) = \int \mathbb{P}^\eta(\eta_t \equiv 1 \text{ on } A) \, \mu(d\eta)$$

$$= \int \mathbb{P}^A(\eta \equiv 1 \text{ on } A_t) \, \mu(d\eta)$$

$$= \int \mathbb{P}^A(\eta \equiv 1 \text{ on } A_t, \, |A_t| > 1) \, \mu(d\eta)$$

$$+ \sum_{y \in \mathbb{Z}^d} \mathbb{P}^A(A_t = \{y\}) \mu(\eta(y) = 1),$$

whence

$$\left| \mu S_t(\{\eta : \eta \equiv 1 \text{ on } A\}) - \alpha \right| \leq 2\mathbb{P}^A(|A_t| > 1).$$

By (10.14), $\mu S_t \Rightarrow (1 - \alpha)\delta_0 + \alpha\delta_1$ as claimed. $\qquad\square$

Partial proof of Theorem 10.12. For $A \in \mathcal{S}$, $A \neq \varnothing$, by Theorem 10.9,

$$(10.15) \qquad \phi_\alpha S_t(\eta \equiv 1 \text{ on } A) = \int \mathbb{P}^\eta(\eta_t \equiv 1 \text{ on } A) \, \phi_\alpha(d\eta)$$

$$= \int \mathbb{P}^A(\eta \equiv 1 \text{ on } A_t) \, \phi_\alpha(d\eta)$$

$$= \mathbb{E}^A(\alpha^{|A_t|}).$$

The quantity $|A_t|$ is non-increasing in t, whence the last expectation converges as $t \to \infty$, by the monotone convergence theorem. Using the inclusion–exclusion principle (as in Exercises 2.2–2.3), we deduce that the μS_t-measure of any cylinder event has a limit, and therefore the weak limit ν_α exists (see the discussion of weak convergence in Section 2.3). Since the initial state ϕ_α is translation-invariant, so is ν_α. We omit the proof of ergodicity, which may be found in [167, 170]. By (10.15) with $A = \{x\}$, $\phi_\alpha S_t(\eta(x) = 1) = \alpha$ for all t, so that $\nu_\alpha(\eta(x) = 1) = \alpha$.

It may be shown that the set $\mathcal{I}$ of invariant measures is exactly the convex hull of the set $\{\nu_\alpha : \alpha \in [0, 1]\}$. The proof of this is omitted, and may be found in [167, 170]. Since the ν_α are ergodic, they are extremal within the class of translation-invariant measures, whence $\mathcal{I}_e = \{\nu_\alpha : \alpha \in [0, 1]\}$. $\qquad\square$

10.4 Exclusion model

In this model for a lattice gas, particles jump around the countable set V, subject to the excluded-volume constraint that no more than one particle may occupy any given vertex at any given time. The state space is $\Sigma = \{0, 1\}^V$, where the local state 1 represents occupancy by a particle. The dynamics are assumed to proceed as follows. Let $P = (p_{x,y} : x, y \in V)$ be the transition matrix of a Markov chain on V. In order to guarantee the existence of the corresponding exclusion process, we shall assume that

$$\sup_{y \in V} \sum_{x \in V} p_{x,y} < \infty.$$

If the current state is $\eta \in \Sigma$, and $\eta(x) = 1$, the particle at x waits an exponentially distributed time, parameter 1, before it attempts to jump. At the end of this holding time, it chooses a vertex y according to the probabilities $p_{x,y}$. If, at this instant, y is empty, then this particle jumps to y. If y is occupied, the jump is suppressed, and the particle remains at x. Particles are deemed to be indistinguishable.

The generator $\mathcal{L}$ of the Markov process is given by

$$\mathcal{L} f(\eta) = \sum_{\substack{x,y \in V: \\ \eta(x)=1, \, \eta(y)=0}} p_{x,y}[f(\eta_{x,y}) - f(\eta)],$$

for cylinder functions f, where $\eta_{x,y}$ is the state obtained from η by interchanging the local states of x and y, that is,

(10.16)
$$\eta_{x,y}(z) = \begin{cases} \eta(x) & \text{if } z = y, \\ \eta(y) & \text{if } z = x, \\ \eta(z) & \text{otherwise.} \end{cases}$$

We may construct the process via a graphical representation, as in Section 10.3. For each $x \in V$, we let Po_x be a Poisson process with rate 1; these are the times at which a particle at x (if, indeed, x is occupied at the relevant time) attempts to move away from x. With each 'time' $T \in \mathrm{Po}_x$, we associate a vertex Y chosen according to the mass function $p_{x,y}$, $y \in V$. If x is occupied by a particle at time T, this particle attempts to jump at this instant of time to the new position Y. The jump is successful if Y is empty at time T, otherwise the move is suppressed.

It is immediate that the two Dirac measures δ_0 and δ_1 are invariant. We shall see below that the family of invariant measures is generally much richer than this. The theory is substantially simpler in the *symmetric* case, and thus we assume henceforth that

(10.17)
$$p_{x,y} = p_{y,x}, \qquad x, y \in V.$$

See [167, Chap. VIII] and [170] for bibliographies for the asymmetric case. If V is the vertex-set of a graph $G = (V, E)$, and P is the transition matrix of simple random walk on G, then (10.17) amounts to the assumption that G be regular.

Mention is made of the *totally asymmetric simple exclusion process* (TASEP), namely the exclusion process on the line $\mathbb{L}$ in which particles may move only in a given direction, say to the right. This apparently simple model has attracted a great deal of attention, and the reader is referred to [86] and the references therein.

We shall see that the exclusion process is self-dual, in the sense of the following Theorem 10.18. Note first that the graphical representation of a symmetric model may be expressed in a slightly simplified manner. For each unordered pair $x, y \in V$, let $\text{Po}_{x,y}$ be a Poisson process with intensity $p_{x,y}$ [$= p_{y,x}$]. For each $T \in \text{Po}_{x,y}$, we interchange the states of x and y at time T. That is, any particle at x moves to y, and vice versa. It is easily seen that the corresponding particle system is the exclusion model. For every $x \in V$, a particle at x at time 0 would pursue a trajectory through V that is determined by the graphical representation, and we denote this trajectory by $R_x(t)$, $t \geq 0$, noting that $R_x(0) = x$. The processes $R_x(\cdot)$, $x \in V$, are of course dependent.

The family $(R_x(\cdot) : x \in V)$ is time-reversible in the following 'strong' sense. Let $t > 0$ be given. For each $y \in V$, we may trace the trajectory arriving at (y, t) backwards in time, and we denote the resulting path by $B_{y,t}(s)$, $0 \leq s \leq t$, with $B_{y,t}(0) = y$. It is clear by the properties of a Poisson process that the families $(R_x(u) : u \in [0, t], x \in V)$ and $(B_{y,t}(s) : s \in [0, t], y \in V)$ have the same laws.

Let $(\eta_t : t \geq 0)$ denote the exclusion model. We distinguish the general model from one possessing only finitely many particles. Let $\mathcal{S}$ be the set of finite subsets of V, and write $(A_t : t \geq 0)$ for an exclusion process with initial state $A_0 \in \mathcal{S}$. We think of η_t as a random $0/1$-vector, and of A_t as a random subset of the vertex-set V.

10.18 Theorem. *Consider a symmetric exclusion model on V. For every $\eta \in \Sigma$ and $A \in \mathcal{S}$,*

(10.19) $\mathbb{P}^\eta(\eta_t \equiv 1 \text{ on } A) = \mathbb{P}^A(\eta \equiv 1 \text{ on } A_t), \qquad t \geq 0.$

Proof. The left side of (10.19) equals the probability that, in the graphical representation: for every $y \in A$, there exists $x \in V$ with $\eta(x) = 1$ such that $R_x(t) = y$. By the remarks above, this equals the probability that $\eta(R_y(t)) = 1$ for every $y \in A$.   □

10.20 Corollary. *Consider a symmetric exclusion model on V. For each $\alpha \in [0, 1]$, the product measure ϕ_α on Σ is invariant.*

Proof. Let η be sampled from Σ according to the product measure ϕ_α. We have that

$$\mathbb{P}^A(\eta \equiv 1 \text{ on } A_t) = \mathbb{E}(\alpha^{|A_t|}) = \alpha^{|A|},$$

since $|A_t| = |A|$. By Theorem 10.18, if η_0 has law ϕ_α, then so does η_t for all t. That is, ϕ_α is an invariant measure. □

The question thus arises of determining the circumstances under which the set of invariant extremal measures is *exactly* the set of product measures.

Assume for simplicity that

(i) $V = \mathbb{Z}^d$,

(ii) the transition probabilities are symmetric and translation-invariant in that

$$p_{x,y} = p_{y,x} = p(y - x), \qquad x, y \in \mathbb{Z}^d,$$

for some function p, and

(iii) the Markov chain with transition matrix $P = (p_{x,y})$ is irreducible.

It can be shown in this case (see [167, 170]) that $\mathcal{I}_e = \{\phi_\alpha : \alpha \in [0, 1]\}$, and that

$$\mu S_t \Rightarrow \phi_\alpha \qquad \text{as } t \to \infty,$$

for any translation-invariant and spatially ergodic probability measure μ with $\mu(\eta(0) = 1) = \alpha$.

In the more general symmetric *non-translation-invariant* case on an arbitrary countable set V, the constants α are replaced by the set $\mathcal{H}$ of functions $\alpha : V \to [0, 1]$ satisfying

(10.21) $$\alpha(x) = \sum_{y \in V} p_{x,y} \alpha(y), \qquad x \in V,$$

that is, the bounded harmonic functions, re-scaled if necessary to take values in $[0, 1]$.[3] Let μ_α be the product measure on Σ with $\mu_\alpha(\eta(x) = 1) = \alpha(x)$. It turns out that the weak limit

$$\nu_\alpha = \lim_{t \to \infty} \mu_\alpha S_t$$

exists, and that $\mathcal{I}_e = \{\nu_\alpha : \alpha \in \mathcal{H}\}$. It may be shown that: ν_α is a product measure if and only if α is a constant function. See [167, 170].

We may find examples in which the set $\mathcal{H}$ is large. Let $P = (p_{x,y})$ be the transition matrix of simple random walk on a binary tree T (each of whose vertices has degree 3, see Figure 6.3). Let 0 be a given vertex of the tree, and think of 0 as the root of three disjoint sub-trees of T. Any solution $(a_n : n \geq 0)$ to the difference equation

(10.22) $$2a_{n+1} - 3a_n + a_{n-1} = 0, \qquad n \geq 1,$$

---

[3] An irreducible symmetric translation-invariant Markov chain on $\mathbb{Z}^d$ has only *constant* bounded harmonic functions. *Exercise*: Prove this statement. It is an easy consequence of the optional stopping theorem for bounded martingales, whenever the chain is recurrent. See [167, pp. 67–70] for a discussion of the general case.

defines a harmonic function α on a given such sub-tree, by $\alpha(x) = a_n$, where n is the distance between 0 and x. The general solution to (10.22) is

$$a_n = A + B(\tfrac{1}{2})^n,$$

where A and B are arbitrary constants. The three pairs (A, B), corresponding to the three sub-trees at 0, may be chosen in an arbitrary manner, subject to the condition that $a_0 = A + B$ is constant across sub-trees. Furthermore, the composite harmonic function on T takes values in $[0, 1]$ if and only if each pair (A, B) satisfies $A, A + B \in [0, 1]$. There exists, therefore, a continuum of admissible non-constant solutions to (10.21), and therefore a continuum of extremal invariant measures of the associated exclusion model.

10.5 Stochastic Ising model

The Ising model is designed as a model of the 'local' interactions of a ferromagnet: each neighbouring pair x, y of vertices have spins contributing $-\sigma_x \sigma_y$ to the energy of the spin-configuration σ. The model is static in time. Physical systems tend to evolve as time passes, and we are thus led to the study of stochastic processes having the Ising model as invariant measure. It is normal to consider Markovian models for time-evolution, and this section contains a very brief summary of some of these. The theory of the dynamics of spin models is very rich, and the reader is referred to [167] and [179, 185, 214] for further introductory accounts.

Let $G = (V, E)$ be a finite connected graph (*infinite* graphs are not considered here). As explained in Section 10.1, a Markov chain on $\Sigma = \{-1, 1\}^V$ is specified by way of its generator $\mathcal{L}$, acting on suitable functions f by

$$(10.23) \qquad \mathcal{L}f(\sigma) = \sum_{\sigma' \in \Sigma} c(\sigma, \sigma')[f(\sigma') - f(\sigma)], \qquad \sigma \in \Sigma,$$

for some function c sometimes called the 'rate (or speed) function'. For $\sigma \neq \sigma'$, we think of $c(\sigma, \sigma')$ as being the rate at which the process jumps to state σ' when currently in state σ. Equation (10.23) requires nothing of the diagonal terms $c(\sigma, \sigma)$, and we choose these such that

$$\sum_{\sigma' \in \Sigma} c(\sigma, \sigma') = 0, \qquad \sigma \in \Sigma.$$

The state space Σ is finite, and thus there is a minimum of technical complications. The probability measure μ is invariant for the process if and only if $\mu \mathcal{L} = 0$, which is to say that

$$(10.24) \qquad \sum_{\sigma \in \Sigma} \mu(\sigma) c(\sigma, \sigma') = 0, \qquad \sigma' \in \Sigma.$$

The process is reversible with respect to μ if and only if the 'detailed balance equations'

(10.25) $$\mu(\sigma)c(\sigma, \sigma') = \mu(\sigma')c(\sigma', \sigma), \qquad \sigma, \sigma' \in \Sigma,$$

hold, in which case μ is automatically invariant.

Let π be the Ising measure on the spin-space Σ satisfying

(10.26) $$\pi(\sigma) \propto e^{-\beta H(\sigma)}, \qquad \sigma \in \Sigma,$$

where $\beta > 0$,

$$H(\sigma) = -h \sum_{x \in V} \sigma_x - \sum_{x \sim y} \sigma_x \sigma_y,$$

and the second summation is over unordered pairs of neighbours. We shall consider Markov processes having π as reversible invariant measure. Many possible choices for the speed function c are possible in (10.25), of which we mention four here.

First, some notation: for $\sigma \in \Sigma$ and $x, y \in V$, the configuration σ_x is obtained from σ by replacing the state of x by $-\sigma(x)$ (see (10.6)), and $\sigma_{x,y}$ is obtained by swapping the states of x and y (see (10.16)). The process is said to proceed by: *spin-flips* if $c(\sigma, \sigma') = 0$ except possibly for pairs σ, σ' that differ on at most one vertex; it proceeds by *spin-swaps* if (for $\sigma \neq \sigma'$) $c(\sigma, \sigma') = 0$ except when $\sigma' = \sigma_{x,y}$ for some $x, y \in V$.

Here are four rate functions that have attracted much attention, presented in a manner that emphasizes applicability to other Gibbs systems. It is easily checked that each is reversible with respect to π.[4]

1. *Metropolis dynamics.* Spin-flip process with

$$c(\sigma, \sigma_x) = \min\left\{1, \exp\left(-\beta[H(\sigma_x) - H(\sigma)]\right)\right\}.$$

2. *Heat-bath dynamics/Gibbs sampler.* Spin-flip process with

$$c(\sigma, \sigma_x) = \left[1 + \exp\left(\beta[H(\sigma_x) - H(\sigma)]\right)\right]^{-1}.$$

This arises as follows. At times of a rate-1 Poisson process, the state at x is replaced by a state chosen at random with the conditional law given $\sigma(y)$, $y \neq x$.

3. *Simple spin-flip dynamics.* Spin-flip process with

$$c(\sigma, \sigma_x) = \exp\left(-\tfrac{1}{2}\beta[H(\sigma_x) - H(\sigma)]\right).$$

4. *Kawasaki dynamics.* Spin-swap process with speed function satisfying

$$c(\sigma, \sigma_{x,y}) = \exp\left(-\tfrac{1}{2}\beta[H(\sigma_{x,y}) - H(\sigma)]\right), \qquad x \sim y.$$

[4]This is Exercise 10.3.

The first three have much in common. In the fourth, Kawasaki dynamics, the 'total magnetization' $M = \sum_x \sigma(x)$ is conserved. This conservation law causes complications in the analysis.

Examples 1–3 are of so-called *Glauber-type*, after Glauber's work on the one-dimensional Ising model, [98]. The term 'Glauber dynamics' is used in several ways in the literature, but may be taken to be a spin-flip process with positive, translation-invariant, finite-range rate function satisfying the detailed balance condition (10.25).

The above dynamics are 'local' in that a transition affects the states of singletons or neighbouring pairs. There is another process called 'Swendsen–Wang dynamics', [228], in which transitions are more extensive. Let π denote the Ising measure (10.26) with $h = 0$. The random-cluster model corresponding to the Ising model with $h = 0$ has state space $\Omega = \{0, 1\}^E$ and parameters $p = 1 - e^{-2\beta}$, $q = 1$. Each step of the Swendsen–Wang evolution comprises two steps: sampling a random-cluster state, followed by resampling a spin configuration. This is made more explicit as follows. Suppose that, at time n, we have obtained a configuration $\sigma_n \in \Sigma$. We construct σ_{n+1} as follows.

I. Let $\omega_n \in \Omega$ be given by: for all $e = \langle x, y \rangle \in E$,

$$\text{if } \sigma_n(x) \neq \sigma_n(y), \quad \text{let } \omega_n(e) = 0,$$

$$\text{if } \sigma_n(x) = \sigma_n(y), \quad \text{let } \omega_n(e) = \begin{cases} 1 & \text{with probability } p, \\ 0 & \text{otherwise,} \end{cases}$$

different edges receiving independent states. The edge-configuration ω_n is carried forward to the next stage.

II. To each cluster C of the graph $(V, \eta(\omega_n))$ we assign an integer chosen uniformly at random from the set $\{1, 2, \ldots, q\}$, different clusters receiving independent labels. Let $\sigma_{n+1}(x)$ be the value thus assigned to the cluster containing the vertex x.

It may be shown that the unique invariant measure of the Markov chain $(\sigma_n : n \geq 1)$ is indeed the Ising measure π. See [109, Sect. 8.5]. Transitions of the Swendsen–Wang algorithm move from a configuration σ to a configuration σ' which is usually very different from σ. Thus, in general, we expect the Swendsen–Wang process to converge faster to equilibrium than the local dynamics given above.

The basic questions for stochastic Ising models concern the rate at which a process converges to its invariant measure, and the manner in which this depends on: (i) the size and topology of G, (ii) any boundary condition that is imposed, and (iii) the values of the external field h and the inverse temperature β. Two ways of quantifying the rate of convergence is via the so-called 'mixing time' and 'relaxation time' of the process. The following discussion is based in part on [18, 165].

Consider a continuous-time Markov process with unique invariant measure μ. The *mixing time* is given as

$$\tau_1 = \inf\left\{t : \sup_{\sigma_1,\sigma_2 \in \Sigma} d_{\mathrm{TV}}(\mathbb{P}_t^{\sigma_1}, \mathbb{P}_t^{\sigma_2}) \le e^{-1}\right\},$$

where

$$d_{\mathrm{TV}}(\mu_1, \mu_2) = \tfrac{1}{2}\sum_{\sigma \in \Sigma}\left|\mu_1(\sigma) - \mu_2(\sigma)\right|,$$

is the *total variation distance* between two probability measures on Σ, and $\mathbb{P}_t^{\sigma}$ denotes the law of the process at time t having started in state σ at time 0.

Write the eigenvalues of the negative generator $-\mathcal{L}$ as

$$0 = \lambda_1 \le \lambda_2 \le \cdots \le \lambda_N.$$

The *relaxation time* τ_2 of the process is defined as the reciprocal of the 'spectral gap' λ_2. It is a general result that

$$\tau_2 \le \tau_1 \le \tau_2\left(1 + \log 1/\left[\min_\sigma \mu(\sigma)\right]\right),$$

so that $\tau_2 \le \tau_1 \le \mathrm{O}(|E|)\tau_2$ for the stochastic Ising model on the connected graph $G = (V, E)$. Therefore, mixing and relaxation times have equivalent orders of magnitude, up to the factor $\mathrm{O}(|E|)$.

No attempt is made here to summarize the very substantial literature on the convergence of Ising models to their equilibria, for which the reader is directed to [185, 214] and more recent works including [179]. A phenomenon of current interest is termed 'cut-off'. It has been observed for certain families of Markov chain that the total variation $d(t) = d_{\mathrm{TV}}(\mathbb{P}_t, \mu)$ has a threshold behaviour: there is a sharp threshold between values of t for which $d(t) \approx 1$, and values for which $d(t) \approx 0$. The relationship between mixing/relaxation times and the cut-off phenomenon is not yet fully understood, but has been studied successfully by Lubetzky and Sly [178] for Glauber dynamics of the high-temperature Ising model in all dimensions.

10.6 Exercises

10.1 [239] *Biased voter model.* Each point of the square lattice is occupied, at each time t, by either a benign or a malignant cell. Benign cells invade their neighbours, each neighbour being invaded at rate β, and similarly malignant cells invade their neighbours at rate μ. Suppose there is exactly one malignant cell at time 0, and let $\kappa = \mu/\beta \geq 1$. Show that the malignant cells die out with probability κ^{-1}.

More generally, what happens on $\mathbb{L}^d$ with $d \geq 2$?

10.2 *Exchangeability.* A probability measure μ on $\{0, 1\}^\mathbb{Z}$ is called *exchangeable* if the quantity $\mu(\{\eta : \eta \equiv 1 \text{ on } A\})$, as A ranges over the set of finite subsets of $\mathbb{Z}$, depends only on the cardinality of A. Show that every exchangeable measure μ is invariant for a symmetric exclusion model on $\mathbb{Z}$.

10.3 *Stochastic Ising model.* Let $\Sigma = \{-1, +1\}^V$ be the state space of a Markov process on the finite graph $G = (V, E)$ which proceeds by spin-flips. The state at $x \in V$ changes value at rate $c(x, \sigma)$ when the state overall is σ. Show that each of the rate functions

$$c_1(x, \sigma) = \min\left\{1, \exp\left(-2\beta \sum_{y \in \partial x} \sigma_x \sigma_y\right)\right\},$$

$$c_2(x, \sigma) = \frac{1}{1 + \exp\left(2\beta \sum_{y \in \partial x} \sigma_x \sigma_y\right)},$$

$$c_3(x, \sigma) = \exp\left(-\beta \sum_{y \in \partial x} \sigma_x \sigma_y\right),$$

gives rise to reversible dynamics with respect to the Ising measure with zero external-field. Here, ∂x denotes the set of neighbours of the vertex x.

11

Random graphs

In the Erdős–Rényi random graph $G_{n,p}$, each pair of vertices is connected by an edge with probability p. We describe the emergence of the giant component when $pn \approx 1$, and identify the density of this component as the survival probability of a Poisson branching process. The Hoeffding inequality may be used to show that, for constant p, the chromatic number of $G_{n,p}$ is asymptotic to $\frac{1}{2}n/\log_\pi n$, where $\pi = 1/(1-p)$.

11.1 Erdős–Rényi graphs

Let $V = \{1, 2, \ldots, n\}$, and let $(X_{i,j} : 1 \le i < j \le n)$ be independent Bernoulli random variables with parameter p. For each pair $i < j$, we place an edge $\langle i, j \rangle$ between vertices i and j if and only if $X_{i,j} = 1$. The resulting random graph is named after Erdős and Rényi [82][1], and it is commonly denoted $G_{n,p}$. The density p of edges may vary with n, for example, $p = \lambda/n$ with $\lambda \in (0, \infty)$, and one commonly considers the structure of $G_{n,p}$ in the limit as $n \to \infty$.

The original motivation for studying $G_{n,p}$ was to understand the properties of 'typical' graphs. This is in contrast to the study of 'extremal' graphs, although it may be noted that random graphs have on occasion manifested properties more extreme than graphs obtained by more constructive means.

Random graphs have proved an important tool in the study of the 'typical' runtime of algorithms. Consider a computational problem associated with graphs, such as the travelling salesman problem. In assessing the speed of an algorithm for this problem, we may find that, in the worst situation, the algorithm is very slow. On the other hand, the typical runtime may be much less than the worst-case runtime. The measurement of 'typical' runtime requires a probability measure on the space of graphs, and it is in this regard that $G_{n,p}$ has risen to prominence within this subfield of theoretical computer science. While $G_{n,p}$ is, in a sense, the obvious candidate for such

[1] See also [96].

205

a probability measure, it suffers from the weakness that the 'mother graph' K_n has a large automorphism group; it is a poor candidate in situations in which pairs of vertices may have differing relationships to one another.

The random graph $G_{n,p}$ has received a very great deal of attention, largely within the community working on probabilistic combinatorics. The theory is based on a mix of combinatorial and probabilistic techniques, and has become very refined.

We may think of $G_{n,p}$ as a percolation model on the complete graph K_n. The parallel with percolation is weak in the sense that the theory of $G_{n,p}$ is largely combinatorial rather than geometrical. There is however a sense in which random graph theory has enriched percolation. The major difficulty in the study of physical systems arises out of the geometry of $\mathbb{R}^d$; points are related to one another in ways that depend greatly on their relative positions in $\mathbb{R}^d$. In a so-called 'mean-field theory', the geometrical component is removed through the assumption that points interact with all other points equally. Mean-field theory leads to an approximate picture of the model in question, and this approximation improves in the limit as $d \to \infty$. The Erdős–Rényi random graph may be seen as a mean-field approximation to percolation. Mean-field models based on $G_{n,p}$ have proved of value for Ising and Potts models also, see [45, 242].

This chapter contains brief introductions to two areas of random-graph theory, each of which uses probability theory in a special way. The first is an analysis of the emergence of the so-called giant component in $G_{n,p}$ with $p = \lambda/n$, as the parameter λ passes through the value $\lambda_c = 1$. Of the several possible ways of doing this, we emphasize here the relevance of arguments from branching processes. The second area considered here is a study of the chromatic number of $G_{n,p}$, as $n \to \infty$ with constant p. This classical problem was solved by Béla Bollobás [42] using Hoeffding's inequality for the tail of a martingale, Theorem 4.21.

The two principal references for the theory of $G_{n,p}$ are the earlier book [44] by Bollobás, and the more recent work [144] of Janson, Łuzcak and Ruciński. We say nothing here about recent developments in random-graph theory involving models for the so-called small world. See [79] for example.

11.2 Giant component

Consider the random graph $G_{n,\lambda/n}$, where $\lambda \in (0, \infty)$ is a constant. We build the component at a given vertex v as follows. The vertex v is adjacent to a certain number N of vertices, where N has the $\mathrm{bin}(n-1, \lambda/n)$ distribution. Each of these vertices is joined to a random number of vertices, distributed approximately as N, and such that, with probability $1 - o(1)$, these new vertex-sets are disjoint. Since the $\text{bin}(n - 1, \lambda/n)$ distribution is 'nearly' Poisson $\text{Po}(\lambda)$, the component at v grows very much like a branching process with family-size distribution $\text{Po}(\lambda)$. The branching-process approximation becomes less good as the component grows, and in particular when its size becomes of order n. The mean family-size equals λ, and thus the process with $\lambda < 1$ is very different from that with $\lambda > 1$.

Suppose that $\lambda < 1$. In this case, the branching process is (almost surely) extinct, and possesses a finite number of vertices. With high probability, the size of the growing cluster at v is sufficiently small to be well approximated by a $\text{Po}(\lambda)$ branching-process. Having built the component at v, we pick another vertex w and act similarly. By iteration, we obtain that $G_{n,p}$ is the union of clusters each with exponentially decaying tail. The largest component has order $\log n$.

When $\lambda > 1$, the branching process grows beyond limits with strictly positive probability. This corresponds to the existence in $G_{n,p}$ of a component of size having order n. We make this more formal as follows. Let X_n be the number of vertices in a largest component of $G_{n,p}$. We write $Z_n = o_p(y_n)$ if $Z_n/y_n \to 0$ in probability as $n \to \infty$. An event A_n is said to occur *asymptotically almost surely* (abbreviated as a.a.s.) if $\mathbb{P}(A_n) \to 1$ as $n \to \infty$.

11.1 Theorem [82]. *We have that*

$$\frac{1}{n} X_n = \begin{cases} o_p(1) & \text{if } \lambda \le 1, \\ \alpha(\lambda)(1 + o_p(1)) & \text{if } \lambda > 1, \end{cases}$$

where $\alpha(\lambda)$ is the survival probability of a branching process with a single progenitor and family-size distribution $\text{Po}(\lambda)$.

It is standard (see [121, Sect. 5.4], for example) that the extinction probability $\eta(\lambda) = 1 - \alpha(\lambda)$ of such a branching process is the smallest non-negative root of the equation $s = G(s)$, where $G(s) = e^{\lambda(s-1)}$. It is left as an exercise[2] to check that

$$\eta(\lambda) = \frac{1}{\lambda} \sum_{k=1}^{\infty} \frac{k^{k-1}}{k!} (\lambda e^{-\lambda})^k.$$

[2]Here is one way that resonates with random graphs. Let p_k be the probability that vertex 1 lies in a component that is a tree of size k. By enumerating the possibilities,

$$p_k = \binom{n-1}{k-1} k^{k-2} \left(\frac{\lambda}{n}\right)^{k-1} \left(1 - \frac{\lambda}{n}\right)^{k(n-k)+\binom{k}{2}-k+1}.$$

Simplify and sum over k.

Proof. By a coupling argument, the distribution of X_n is non-decreasing in λ. Since $\alpha(1) = 0$, it suffices to consider the case $\lambda > 1$, and we assume this henceforth. We follow [144, Sect. 5.2], and use a branching-process argument. (See also [21].) Choose a vertex v. At the first step, we find all neighbours of v, say $v_1, v_2, \ldots, v_r$, and we mark v as *dead*. At the second step, we generate all neighbours of v_1 in $V \setminus \{v, v_1, v_2, \ldots, v_r\}$, and we mark v_1 as dead. This process is iterated until the entire component of $G_{n,p}$ containing v has been generated. Any vertex thus discovered in the component of v, but not yet dead, is said to be *live*. Step i is said to be complete when there are exactly i dead vertices. The process terminates when there are no live vertices remaining.

Conditional on the history of the process up to and including the $(i-1)$th step, the number N_i of vertices added at step i is distributed as $\mathrm{bin}(n-m, p)$, where m is the number of vertices already generated.

Let
$$k_- = \frac{16\lambda}{(\lambda - 1)^2} \log n, \quad k_+ = n^{2/3}.$$

In this section, all logarithms are natural. Consider the above process started at v, and let A_v be the event that: either the process terminates after fewer than k_- steps, or, for every k satisfying $k_- \le k \le k_+$, there are at least $\frac{1}{2}(\lambda - 1)k$ live vertices after step k. If A_v does not occur, there exists $k \in [k_-, k_+]$ such that: step k takes place and, after its completion, fewer than
$$m = k + \tfrac{1}{2}(\lambda - 1)k = \tfrac{1}{2}(\lambda + 1)k$$

vertices have been discovered in all. For simplicity of notation, we assume that $\frac{1}{2}(\lambda + 1)k$ is an integer.

On the event $\overline{A}_v$, and with such a choice for k,
$$(N_1, N_2, \ldots, N_k) \ge_{\mathrm{st}} (Y_1, Y_2, \ldots, Y_k),$$

where the Y_j are independent random variables with the binomial distribution[3] $\mathrm{bin}(n - \frac{1}{2}(\lambda + 1)k, p)$. Therefore,
$$1 - \mathbb{P}(A_v) \le \sum_{k=k_-}^{k_+} \pi_k,$$

where

(11.2) $$\pi_k = \mathbb{P}\left(\sum_{i=1}^{k} Y_i \le \tfrac{1}{2}(\lambda + 1)k \right).$$

---

[3]Here and later, we occasionally use fractions where integers are required.

Now, $Y_1 + Y_2 + \cdots + Y_k$ has the $\mathrm{bin}(k(n - \frac{1}{2}(\lambda+1)k), p)$ distribution. By the Chernoff bound[4] for the tail of the binomial distribution, for $k_- \leq k \leq k_+$ and large n,

$$\pi_k \leq \exp\left(-\frac{(\lambda-1)^2 k^2}{9\lambda k}\right) \leq \exp\left(-\frac{(\lambda-1)^2}{9\lambda}k_-\right)$$
$$= O(n^{-16/9}).$$

Therefore, $1 - \mathbb{P}(A_v) \leq k_+ O(n^{-16/9}) = o(n^{-1})$, and this proves that

$$\mathbb{P}\left(\bigcap_{v \in V} A_v\right) \geq 1 - \sum_{v \in V}[1 - \mathbb{P}(A_v)] \to 1 \qquad \text{as } n \to \infty.$$

In particular, a.a.s., no component of $G_{n,\lambda/n}$ has size between k_- and k_+.

We show next that, a.a.s., there do not exist more than two components with size exceeding k_+. Assume that $\bigcap_v A_v$ occurs, and let v', v'' be distinct vertices lying in components with size exceeding k_+. We run the above process beginning at v' for the first k_+ steps, and we finish with a set L' containing at least $\frac{1}{2}(\lambda - 1)k_+$ live vertices. We do the same for the process from v''. Either the growing component at v'' intersects the current component v' by step k_+, or not. If the latter, then we finish with a set L'', containing at least $\frac{1}{2}(\lambda - 1)k_+$ live vertices, and disjoint from L'. The chance (conditional on arriving at this stage) that there exists no edge between L' and L'' is bounded above by

$$(1-p)^{\lfloor \frac{1}{2}(\lambda-1)k_+ \rfloor^2} \leq \exp\left(-\tfrac{1}{4}\lambda(\lambda-1)^2 n^{1/3}\right) = o(n^{-2}).$$

Therefore, the probability that there exist two distinct vertices belonging to distinct components of size exceeding k_+ is no greater than

$$1 - \mathbb{P}\left(\bigcap_{v \in V} A_v\right) + n^2 o(n^{-2}) = o(1).$$

In summary, a.a.s., every component is either 'small' (smaller than k_-) or 'large' (larger than k_+), and there can be no more than one large component. In order to estimate the size of any such large component, we use Chebyshev's inequality to estimate the aggregate sizes of small components. Let $v \in V$. The chance $\sigma = \sigma(n, p)$ that v is in a small component satisfies

(11.3) $$\eta_- - o(1) \leq \sigma \leq \eta_+,$$

where η_+ (respectively, η_-) is the extinction probability of a branching process with family-size distribution $\mathrm{bin}(n - k_-, p)$ (respectively, $\mathrm{bin}(n, p)$), and the o(1) term bounds the probability that the latter branching process terminates after k_- or more steps. It is an easy exercise[5] to show that

---

[4]See Exercise 11.3.

[5]See Exercise 11.2.

$\eta_-, \eta_+ \to \eta$ as $n \to \infty$, where $\eta(\lambda) = 1 - \alpha(\lambda)$ is the extinction probability of a $\mathrm{Po}(\lambda)$ branching process.

The number S of vertices in small components satisfies

$$\mathbb{E}(S) = \sigma n = (1 + \mathrm{o}(1))\eta n.$$

Furthermore, by an argument similar to that above,

$$\mathbb{E}(S(S-1)) \le n\sigma\big[k_- + n\sigma(n - k_-, p)\big] = (1 + \mathrm{o}(1))(\mathbb{E}S)^2,$$

whence, by Chebyshev's inequality, $G_{n,p}$ possesses $(\eta + \mathrm{o}_p(1))n$ vertices in small components. This leaves just $n - (\eta + \mathrm{o}_p(1))n = (\alpha + \mathrm{o}_p(1))n$ vertices remaining for the large component, and the theorem is proved. $\quad\square$

A further analysis yields the size X_n of the largest subcritical component, and the size Y_n of the second largest supercritical component.

11.4 Theorem.

(a) *When $\lambda < 1$,*

$$X_n = (1 + \mathrm{o}_p(1))\frac{\log n}{\lambda - 1 - \log \lambda}.$$

(b) *When $\lambda > 1$,*

$$Y_n = (1 + \mathrm{o}_p(1))\frac{\log n}{\lambda' - 1 - \log \lambda'},$$

where $\lambda' = \lambda(1 - \alpha(\lambda))$.

If $\lambda > 1$, and we remove the largest component, we are left with a random graph on $n - X_n \sim n(1 - \alpha(\lambda))$ vertices. The mean vertex-degree of this subgraph is approximately

$$\frac{\lambda}{n} \cdot n(1 - \alpha(\lambda)) = \lambda(1 - \alpha(\lambda)) = \lambda'.$$

It may be checked that this is strictly smaller than 1, implying that the remaining subgraph behaves as a subcritical random graph on $n - X_n$ vertices. Theorem 11.4(b) now follows from part (a).

The picture is more interesting when $\lambda \approx 1$, for which there is a detailed combinatorial study of [143]. Rather than describing this here, we deviate to the work of David Aldous [16], who has demonstrated a link, via the multiplicative coalescent, to Brownian motion. We set

$$p = \frac{1}{n} + \frac{t}{n^{4/3}},$$

where $t \in \mathbb{R}$, and we write $C_n^t(1) \ge C_n^t(2) \ge \cdots$ for the component sizes of $G_{n,p}$ in decreasing order. We shall explore the weak limit (as $n \to \infty$) of the sequence $n^{-2/3}(C_n^t(1), C_n^t(2), \dots)$.

Let $W = (W(s) : s \geq 0)$ be a standard Brownian motion, and

$$W^t(s) = W(s) + ts - \tfrac{1}{2}s^2, \qquad s \geq 0,$$

a Brownian motion with drift $t - s$ at time s. Write

$$B^t(s) = W^t(s) - \inf_{0 \leq s' \leq s} W^t(s')$$

for a reflecting inhomogenous Brownian motion with drift.

11.5 Theorem [16]. *As $n \to \infty$,*

$$n^{-2/3}(C_n^t(1), C_n^t(2), \dots) \Rightarrow (C^t(1), C^t(2), \dots),$$

where $C^t(j)$ is the length of the jth largest excursion of B^t.

We think of the sequences of Theorem 11.5 as being chosen at random from the space of decreasing non-negative sequences $\mathbf{x} = (x_1, x_2, \dots)$, with metric

$$d(\mathbf{x}, \mathbf{y}) = \sqrt{\sum_i (x_i - y_i)^2}.$$

As t increases, two components of sizes x_i, x_j 'coalesce' at a rate proportional to the product $x_i x_j$. Theorem 11.5 identifies the scaling limit of this process as that of the evolving excursion-lengths of W^t reflected at zero. This observation has contributed to the construction of the so-called 'multiplicative coalescent'.

In summary, the largest component of the subcritical random graph (when $\lambda < 1$) has order $\log n$, and of the supercritical graph (when $\lambda > 1$) order n. When $\lambda = 1$, the largest component has order $n^{2/3}$, with a multiplicative constant that is a random variable. The discontinuity at $\lambda = 1$ is sometimes referred to as the 'Erdős–Rényi double jump'.

11.3 Independence and colouring

Our second random-graph study is concerned with the chromatic number of $G_{n,p}$ for constant p. The theory of graph-colourings is a significant part of graph theory. The *chromatic number* $\chi(G)$ of a graph G is the least number of colours with the property that: there exists an allocation of colours to vertices such that no two neighbours have the same colour. Let $p \in (0, 1)$, and write $\chi_{n,p}$ for the chromatic number of $G_{n,p}$.

A subset W of V is called *independent* if no pair of vertices in W are adjacent, that is, if $X_{i,j} = 0$ for all $i, j \in W$. Any colouring of $G_{n,p}$ partitions V into independent sets each with a given colour, and therefore the chromatic number is related to the size $I_{n,p}$ of the largest independent set of $G_{n,p}$.

11.6 Theorem [116]. *We have that*

$$I_{n,p} = (1 + o_p(1))2\log_\pi n,$$

where the base π of the logarithm is $\pi = 1/(1-p)$.

The proof follows a standard route: the upper bound follows by an estimate of an expectation, and the lower by an estimate of a second moment. When performed with greater care, such calculations yield much more accurate estimates of $I_{n,p}$ than those presented here, see, for example, [44], [144, Sect. 7.1], and [186, Sect. 2]. Specifically, there exists an integer-valued function $r = r(n, p)$ such that

$$(11.7) \qquad \mathbb{P}(r - 1 \le I_{n,p} \le r) \to 1 \qquad \text{as } n \to \infty.$$

Proof. Let N_k be the number of independent subsets of V with cardinality k. Then

$$(11.8) \qquad \mathbb{P}(I_{n,p} \ge k) = \mathbb{P}(N_k \ge 1) \le \mathbb{E}(N_k).$$

Now,

$$(11.9) \qquad \mathbb{E}(N_k) = \binom{n}{k}(1-p)^{\binom{k}{2}},$$

With $\epsilon > 0$, set $k = 2(1 + \epsilon)\log_\pi n$, and use the fact that

$$\binom{n}{k} \le \frac{n^k}{k!} \le (ne/k)^k,$$

to obtain

$$\log_\pi \mathbb{E}(N_k) \le -(1 + o(1))k\epsilon \log_\pi n \to -\infty \qquad \text{as } n \to \infty.$$

By (11.8), $\mathbb{P}(I_{n,p} \ge k) \to 0$ as $n \to \infty$. This is an example of the use of the so-called 'first-moment method'.

A lower bound for $I_{n,p}$ is obtained by the 'second-moment method' as follows. By Chebyshev's inequality,

$$\mathbb{P}(N_k = 0) \le \mathbb{P}(|N_k - \mathbb{E}N_k| \ge \mathbb{E}N_k) \le \frac{\text{var}(N_k)}{\mathbb{E}(N_k)^2},$$

whence, since N_k takes values in the non-negative integers,

$$(11.10) \qquad \mathbb{P}(N_k \ge 1) \ge 2 - \frac{\mathbb{E}(N_k^2)}{\mathbb{E}(N_k)^2}.$$

Let $\epsilon > 0$ and $k = 2(1 - \epsilon)\log_\pi n$. By (11.10), it suffices to show that

$$(11.11) \qquad \frac{\mathbb{E}(N_k^2)}{\mathbb{E}(N_k)^2} \to 1 \qquad \text{as } n \to \infty.$$

By the Cauchy–Schwarz inequality, the left side is at least 1. By an elementary counting argument,

$$\mathbb{E}(N_k^2) = \binom{n}{k}(1-p)^{\binom{k}{2}} \sum_{i=0}^{k} \binom{k}{i}\binom{n-k}{k-i}(1-p)^{\binom{k}{2}-\binom{i}{2}}.$$

After a minor analysis using (11.9) and (11.11), we may conclude that $\mathbb{P}(I_{n,p} \geq k) \to 1$ as $n \to \infty$. The theorem is proved. $\qquad\square$

We turn now to the chromatic number $\chi_{n,p}$. Since the size of any set of vertices of given colour is no larger than $I_{n,p}$, we have immediately that

$$(11.12) \qquad \chi_{n,p} \geq \frac{n}{I_{n,p}} = (1 + o_p(1))\frac{n}{2\log_\pi n}.$$

The sharpness of this inequality was proved by Béla Bollobás [42], in a striking application of Hoeffding's inequality for the tail of a martingale, Theorem 4.21.

11.13 Theorem [42]. *We have that*

$$\chi_{n,p} = (1 + o_p(1))\frac{n}{2\log_\pi n},$$

where $\pi = 1/(1-p)$.

The term $o_p(1)$ may be estimated quite precisely by a more detailed analysis than that presented here, see [42, 187] and [144, Sect. 7.3]. Specifically, we have, a.a.s., that

$$\chi_{n,p} = \frac{n}{2\log_\pi n - 2\log_\pi \log_\pi n + O_p(1)},$$

where $Z_n = O_p(y_n)$ means $\mathbb{P}(|Z_n/y_n| > M) \leq g(M) \to 0$ as $M \to \infty$.

Proof. The lower bound follows as in (11.12), and so we concentrate on finding an upper bound for $\chi_{n,p}$. Let $0 < \epsilon < \frac{1}{4}$, and write

$$k = \lfloor 2(1-\epsilon)\log_\pi n \rfloor.$$

We claim that, with probability $1 - o(1)$, *every* subset of V with cardinality at least $m = \lfloor n/(\log_\pi n)^2 \rfloor$ possesses an independent subset of size at least k. The required bound on $\chi_{n,p}$ follows from this claim, as follows. We find an independent set of size k, and we colour its vertices with colour 1. From the remaining set of $n - k$ vertices, we find an independent set of size k, and we colour it with colour 2. This process may be iterated until there remains a set S of size smaller than $\lfloor n/(\log_\pi n)^2 \rfloor$. We colour the vertices of S 'greedily', using $|S|$ further colours. The total number of colours used in the above algorithm is no greater than

$$\frac{n}{k} + \frac{n}{(\log_\pi n)^2},$$

which, for large n, is smaller than $\frac{1}{2}(1 + 2\epsilon)n/\log_\pi n$. The required claim is a consequence of the following lemma.

11.14 Lemma. *With the above notation, the probability that $G_{m,p}$ contains no independent set of size k is less than* $\exp\left(-n^{\frac{7}{2}-2\epsilon+o(1)}/m^2\right)$.

There are $\binom{n}{m}$ $(< 2^n)$ subsets of $\{1, 2, \ldots, n\}$ with cardinality m. The probability that some such subset fails to contain an independent set of size k is, by the lemma, no larger than

$$2^n \exp(-n^{\frac{7}{2}-2\epsilon+o(1)}/m^2) = o(1).$$

We turn to the proof of Lemma 11.14, for which we shall use the Hoeffding inequality, Theorem 4.21.

For $M \geq k$, let

$$(11.15) \qquad\qquad F(M, k) = \binom{M}{k}(1 - p)^{\binom{k}{2}}.$$

We shall require M to be such that $F(M, k)$ grows in the manner of a power of n, and to that end we set

$$(11.16) \qquad\qquad M = \lfloor (Ck/e)n^{1-\epsilon} \rfloor,$$

where

$$\log_\pi C = \frac{3}{8(1 - \epsilon)}$$

has been chosen in such a way that

$$(11.17) \qquad\qquad F(M, k) = n^{\frac{7}{4}-\epsilon+o(1)}.$$

Let $\mathcal{I}(r)$ be the set of independent subsets of $\{1, 2, \ldots, r\}$ with cardinality k. We write $N_k = |\mathcal{I}(m)|$, and N_k' for the number of elements I of $\mathcal{I}(m)$ with the property that $|I \cap I'| \leq 1$ for all $I' \in \mathcal{I}(m)$, $I' \neq I$. Note that

$$(11.18) \qquad\qquad N_k \geq N_k'.$$

We shall estimate $\mathbb{P}(N_k = 0)$ by applying Hoeffding's inequality to a martingale constructed in a standard manner from the random variable N_k'. First, we order as $(e_1, e_2, \ldots, e_{\binom{m}{2}})$ the edges of the complete graph on the vertex-set $\{1, 2, \ldots, m\}$. Let $\mathcal{F}_s$ be the σ-field generated by the states of the edges $e_1, e_2, \ldots, e_s$, and let $Y_s = \mathbb{E}(N'_k \mid \mathcal{F}_s)$. It is elementary that the sequence $(Y_s, \mathcal{F}_s)$, $0 \le s \le \binom{m}{2}$, is a martingale (see [121, Example 7.9.24]). The quantity N'_k has been defined in such a way that the addition or removal of an edge causes its value to change by at most 1. Therefore, the martingale differences satisfy $|Y_{s+1} - Y_s| \le 1$. Since $Y_0 = \mathbb{E}(N'_k)$ and $Y_{\binom{m}{2}} = N'_k$,

$$
\begin{align}
(11.19) \qquad \mathbb{P}(N_k = 0) &\le \mathbb{P}(N'_k = 0) \\
&= \mathbb{P}\big(N'_k - \mathbb{E}(N'_k) \le -\mathbb{E}(N'_k)\big) \\
&\le \exp\left(-\tfrac{1}{2}\mathbb{E}(N'_k)^2 \middle/ \binom{m}{2}\right) \\
&\le \exp\big(-\mathbb{E}(N'_k)^2/m^2\big),
\end{align}
$$

by (11.18) and Theorem 4.21. We now require a lower bound for $\mathbb{E}(N'_k)$.

Let M be as in (11.16). Let $M_k = |\mathcal{I}(M)|$, and let M'_k be the number of elements $I \in \mathcal{I}(M)$ such that $|I \cap I'| \le 1$ for all $I' \in \mathcal{I}(M)$, $I' \ne I$. Since $m \ge M$,

$$
(11.20) \qquad N'_k \ge M'_k,
$$

and we shall bound $\mathbb{E}(M'_k)$ from below. Let $K = \{1, 2, \ldots, k\}$, and let A be the event that K is an independent set. Let Z be the number of elements of $\mathcal{I}(M)$, other than K, that intersect K in two or more vertices. Then

$$
\begin{align}
(11.21) \qquad \mathbb{E}(M'_k) &= \binom{M}{k} \mathbb{P}(A \cap \{Z = 0\}) \\
&= \binom{M}{k} P(A)\mathbb{P}(Z = 0 \mid A) \\
&= F(M, k)\mathbb{P}(Z = 0 \mid A).
\end{align}
$$

We bound $\mathbb{P}(Z = 0 \mid A)$ by

$$
\begin{align}
(11.22) \quad \mathbb{P}(Z = 0 \mid A) &= 1 - \mathbb{P}(Z \ge 1 \mid A) \\
&\ge 1 - \mathbb{E}(Z \mid A) \\
&= 1 - \sum_{t=2}^{k-1} \binom{k}{t}\binom{M-k}{k-t}(1 - p)^{\binom{k}{2}-\binom{t}{2}} \\
&= 1 - \sum_{t=2}^{k-1} F_t, \qquad \text{say.}
\end{align}
$$

For $t \geq 2$,

(11.23)
$$F_t/F_2 = \frac{(M - 2k + 2)!}{(M - 2k + t)!} \cdot \left(\frac{(k-2)!}{(k-t)!}\right)^2 \cdot \frac{2}{t!}(1-p)^{-\frac{1}{2}(t+1)(t-2)}$$
$$\leq \left[\frac{k^2(1-p)^{-\frac{1}{2}(t+1)}}{M - 2k}\right]^{t-2}.$$

For $2 \leq t \leq \frac{1}{2}k$,

$$\log_\pi\left[(1-p)^{-\frac{1}{2}(t+1)}\right] \leq \tfrac{1}{4}(k+2) \leq \tfrac{1}{2} + \tfrac{1}{2}(1-\epsilon)\log_\pi n,$$

so $(1-p)^{-\frac{1}{2}(t+1)} = O(n^{\frac{1}{2}(1-\epsilon)})$. By (11.23),

$$\sum_{2 \leq t \leq \frac{1}{2}k} F_t = (1 + o(1))F_2.$$

Similarly,

$$F_t/F_{k-1} = \binom{k}{t}\binom{M-k}{k-t}\frac{(1-p)^{\frac{1}{2}(k+t-2)(k-t-1)}}{k(M-k)}$$
$$\leq \left[kn(1-p)^{\frac{1}{2}(k+t-2)}\right]^{k-t-1}.$$

For $\frac{1}{2}k \leq t \leq k - 1$, we have as above that

$$(1-p)^{\frac{1}{2}(k+t)} \leq (1-p)^{\frac{3}{4}k} \leq n^{-\frac{9}{8}},$$

whence

$$\sum_{\frac{1}{2}k < t \leq k-1} F_t = (1 + o(1))F_{k-1}.$$

In summary,

(11.24)
$$\sum_{t=2}^{k-1} F_t = (1 + o(1))(F_2 + F_{k-1}).$$

By (11.15) and (11.17),

$$F_2 \leq \frac{k^4}{2(1-p)(M-k)^2}F(M, k)$$
$$= n^{-\frac{1}{4}+\epsilon+o(1)} = o(1),$$

and similarly

$$F_{k-1} = k(M-k)(1-p)^{k-1} = o(1).$$

By (11.24) and (11.21)–(11.22),

$$\mathbb{E}(M_k') = (1 + o(1))F(M, k) = n^{\frac{7}{4} - \epsilon + o(1)}.$$

Returning to the martingale bound (11.19), it follows by (11.20) that

$$\mathbb{P}(N_k = 0) \le \exp(-n^{\frac{7}{2} - 2\epsilon + o(1)}/m^2)$$

as required. $\qquad\qquad\qquad\qquad\qquad\qquad\qquad\qquad\qquad\qquad\qquad\qquad$ □

11.4 Exercises

11.1 Let $\eta(\lambda)$ be the extinction probability of a branching process whose family-sizes have the Poisson distribution Po(λ). Show that

$$\eta(\lambda) = \frac{1}{\lambda} \sum_{k=1}^{\infty} \frac{k^{k-1}}{k!} (\lambda e^{-\lambda})^k.$$

11.2 Consider a branching process whose family-sizes have the binomial distribution bin($n, \lambda/n$). Show that the extinction probability converges to $\eta(\lambda)$ as $n \to \infty$, where $\eta(\lambda)$ is the extinction probability of a branching process with family-sizes distributed as Po(λ).

11.3 *Chernoff bounds.* Let X have the binomial distribution bin(n, p), and let $\lambda = np$. Obtain exponentially decaying bounds for the probabilities of upper and lower deviations of X from its mean λ, such as those to be found in [144, Sect. 2.1]:

$$\mathbb{P}(X \ge \lambda + t) \le e^{-\lambda\phi(t/\lambda)} \le \exp\left(-\frac{t^2}{2(\lambda + t/3)}\right), \quad t \ge 0,$$

$$\mathbb{P}(X \le \lambda - t) \le e^{-\lambda\phi(-t/\lambda)} \le \exp\left(-\frac{t^2}{2\lambda}\right), \qquad t \ge 0,$$

where

$$\phi(x) = \begin{cases} (1 + x)\log(1 + x) - x & \text{if } x \ge -1, \\ \infty & \text{if } x < -1. \end{cases}$$

11.4 [44] Show that the size of the largest independent set of $G_{n,p}$ is, a.a.s., either $r - 1$ or r, for some deterministic function $r = r(n, p)$.

11.5 Consider a branching process with a single progenitor and family-sizes distributed as the random variable X. Let

$$T = \min\{n \ge 1 : X_1 + X_2 + \cdots + X_n = n - 1\},$$

where the X_i are independent copies of X. Show that T has the same distribution as the total number of individuals in the branching process. (The minimum of the empty set is defined to be ∞.)

11.6 (continuation) In the random graph $G_{n,p}$ with $p = \lambda/n$, where $\lambda \in (0, 1)$, show that the size M_n of the largest cluster satisfies $\mathbb{P}(M_n \geq a \log n) \to 0$ as $n \to \infty$ for any $a > \lambda - 1 - \log \lambda$.

11.7 (continuation) Prove the complementary fact that $\mathbb{P}(M_n \leq a \log n) \to 0$ as $n \to \infty$ for any $a < \lambda - 1 - \log \lambda$.

12

Lorentz gas

A small particle is fired through an environment of large particles, and is subjected to reflections on impact. Little is known about the trajectory of the small particle when the larger ones are distributed at random. The notorious problem on the square lattice is summarized, and open questions are posed for the case of a continuum of needle-like mirrors in the plane.

12.1 Lorentz model

In a famous sequence [176] of papers of 1906, Hendrik Lorentz introduced a version of the following problem. Large (heavy) particles are distributed about $\mathbb{R}^d$. A small (light) particle is fired through $\mathbb{R}^d$, with a trajectory comprising straight-line segments between the points of interaction with the heavy particles. When the small particle hits a heavy particle, the small particle is reflected at its surface, and the large particle remains motionless. See Figure 12.1 for an illustration.

We may think of the heavy particles as objects bounded by reflecting surfaces, and the light particle as a photon. The problem is to say something non-trivial about how the trajectory of the photon depends on the 'environment' of heavy particles. Conditional on the environment, the photon pursues a deterministic path about which the natural questions include:

1. Is the path unbounded?
2. How distant is the photon from its starting point after time t?

For simplicity, we assume henceforth that the large particles are identical to one another, and that the small particle has negligible volume.

Probability may be injected naturally into this model through the assumption that the heavy particles are distributed at random around $\mathbb{R}^d$ according to some probability measure μ. The questions above may be rephrased, and made more precise, in the language of probability theory. Let X_t denote the position of the photon at time t, assuming constant velocity. Under what conditions on μ:

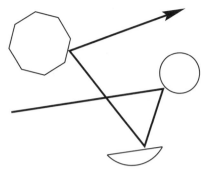

Figure 12.1. The trajectory of the photon comprises straight-line segments between the points of reflection.

I. Is there strictly positive probability that the function X_t is unbounded?

II. Does X_t converge to a Brownian motion, after suitable re-scaling?

For a wide choice of measures μ, these questions are currently unanswered.

The Lorentz gas is very challenging to mathematicians, and little is known
rigorously in reply to the questions above. The reason is that, as the photon
moves around space, it gathers information about the random environment,
and it carries this information with it for ever more.

The Lorentz gas was developed by Paul Ehrenfest [81]. For the relevant
references in the mathematics and physics journals, the reader is referred to
[106, 107]. Many references may be found in [229].

12.2 The square Lorentz gas

Probably the most provocative version of the Lorentz gas for probabilists
arises when the light ray is confined to the square lattice $\mathbb{L}^2$. At each vertex
v of $\mathbb{L}^2$, we place a 'reflector' with probability p, and nothing otherwise
(the occupancies of different vertices are independent). Reflectors come in
two types: 'NW' and 'NE'. A NW reflector deflects incoming rays heading
northwards (respectively, southwards) to the west (respectively, east) and
vice versa. NE reflectors behave similarly with east and west interchanged.
See Figure 12.2. We think of a reflector as being a two-sided mirror placed
at $45°$ to the axes, so that an incoming light ray is reflected along an axis
perpendicular to its direction of arrival. Now, for each vertex v, with probability p we place a reflector at v, and otherwise we place nothing at v. This
is done independently for different v. If a reflector is placed at v, then we
specify that it is equally likely to be NW as NE.

We shine a torch northwards from the origin. The light is reflected by

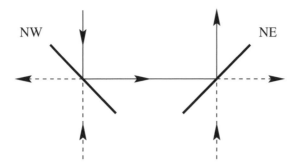

Figure 12.2. An illustration of the effects of NW and NE reflectors on the light ray.

the mirrors, and it is easy to see that: either the light ray is unbounded, or it traverses a closed loop of $\mathbb{L}^2$ (generally with self-crossings). Let

$$\eta(p) = \mathbb{P}_p(\text{the light ray returns to the origin}).$$

Very little is known about the function η. It seems reasonable to conjecture that η is non-decreasing in p, but this has not been proved. If $\eta(p) = 1$, the light follows (almost surely) a closed loop, and we ask: for which p does $\eta(p) = 1$? Certainly, $\eta(0) = 0$, and it is well known that $\eta(1) = 1$.[1]

12.1 Theorem. *We have that $\eta(1) = 1$.*

We invite the reader to consider whether or not $\eta(p) = 1$ for some $p \in (0, 1)$. A variety of related conjectures, not entirely self-consistent, may be found in the physics literature. There are almost no mathematical results about this process beyond Theorem 12.1. We mention the paper [204], where it is proved that the number $N(p)$ of unbounded light rays on $\mathbb{Z}^2$ is almost surely constant, and is equal to one of $0, 1, \infty$. Furthermore, if there exist unbounded light trajectories, then they self-intersect infinitely often. If $N(p) = \infty$, the position X_n of the photon at time n, when following an unbounded trajectory, is superdiffusive in the sense that $\mathbb{E}(|X_n|^2)/n$ is unbounded as $n \to \infty$. The principal method of [204] is to observe the environment of mirrors as viewed from the moving photon.

In a variant of the standard random walk, termed the 'burn-your-bridges' random walk by Omer Angel, an edge is destroyed immediately after it is traversed for the first time. When $p = \frac{1}{3}$, the photon follows a burn-your-bridges random walk on $\mathbb{L}^2$.

---

[1] See the historical remark in [105].

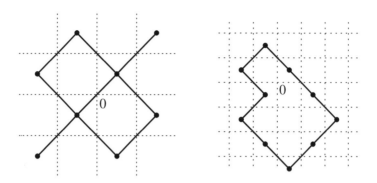

Figure 12.3.   (a) The heavy lines form the lattice $\mathcal{L}$, and the central
point is the origin of $\mathbb{L}^2$. (b) An open cycle in $\mathcal{L}$ constitutes a barrier of
mirrors through which no light may penetrate.

Proof. We construct an ancillary lattice $\mathcal{L}$ as follows. Let

$$A = \left\{ (m + \tfrac{1}{2}, n + \tfrac{1}{2}) : m + n \text{ is even} \right\}.$$

Let $\sim$ be the adjacency relation on A given by $(m + \tfrac{1}{2}, n + \tfrac{1}{2}) \sim (r + \tfrac{1}{2}, s + \tfrac{1}{2})$
if and only if $|m - r| = |n - s| = 1$. We obtain thus a graph $\mathcal{L}$ on A that is
isomorphic to $\mathbb{L}^2$. See Figure 12.3.

We declare the edge of $\mathcal{L}$ joining $(m - \tfrac{1}{2}, n - \tfrac{1}{2})$ to $(m + \tfrac{1}{2}, n + \tfrac{1}{2})$ to
be *open* if there is a NE mirror at (m, n); similarly, we declare the edge
joining $(m - \tfrac{1}{2}, n + \tfrac{1}{2})$ to $(m + \tfrac{1}{2}, n - \tfrac{1}{2})$ to be *open* if there is a NW mirror
at (m, n). Edges that are not open are designated *closed*. This defines a
bond percolation process in which north-easterly and north-westerly edges
are open with probability $\tfrac{1}{2}$. Since $p_c(\mathbb{L}^2) = \tfrac{1}{2}$, the process is critical, and
the percolation probability θ satisfies $\theta(\tfrac{1}{2}) = 0$. See Sections 5.5–5.6.

Let N be the number of open cycles in $\mathcal{L}$ with the origin in their interiors.
Since there is (a.s.) no infinite cluster in the percolation process on the dual
lattice, we have that $\mathbb{P}(N \geq 1) = 1$. Such an open cycle corresponds to a
barrier of mirrors surrounding the origin (see Figure 12.3), from which no
light can escape. Therefore $\eta(1) = 1$.   □

The problem above may be stated for other lattices such as $\mathbb{L}^d$, see [105]
for example. It is much simplified if we allow the photon to flip its own
coin as it proceeds through the disordered medium of mirrors. Two such
models have been explored. In the first, there is a positive probability that
the photon will misbehave when it hits a mirror, see [234]. In the second,
there is allowed a small density of vertices at which the photon acts in the
manner of a random walk, see [38, 117].

12.3 In the plane

Here is a continuum version of the Lorentz gas. Let Π be a Poisson process in $\mathbb{R}^2$ with intensity 1. For each $x \in \Pi$, we place a needle (that is, a closed rectilinear line-segment) of given length l with its centre at x. The orientations of the needles are taken to be independent random variables with a common law μ on $[0, \pi)$. We call μ *degenerate* if it has support on a singleton, that is, if all needles are (almost surely) parallel.

Each needle is interpreted as a (two-sided) reflector of light. Needles are permitted to overlap. Light is projected from the origin northwards, and deflected by the needles. Since the light strikes the endpoint of some needle with probability 0, we shall overlook this possibility.

In a related problem, we may study the union M of the needles, viewed as subsets of $\mathbb{R}^2$, and ask whether either (or both) of the sets $M, \mathbb{R}^2 \setminus M$ contains an unbounded component. This problem is known as 'needle percolation', and it has received some attention (see, for example, [188, Sect. 8.5], and also [134]). Of concern to us in the present setting is the following. Let $\lambda(l) = \lambda_\mu(l)$ be the probability that there exists an unbounded path of $\mathbb{R}^2 \setminus M$ with the origin 0 as endpoint. It is clear that $\lambda(l)$ is non-increasing in l. The following is a fairly straightforward exercise of percolation type.

12.2 Theorem [134]. *There exists $l_c = l_c(\mu) \in (0, \infty]$ such that*

$$\lambda(l) \begin{cases} > 0 & \text{if } l < l_c, \\ = 0 & \text{if } l > l_c, \end{cases}$$

and furthermore $l_c < \infty$ if and only if μ is non-degenerate.

The phase transition has been defined here in terms of the existence of an unbounded 'vacant path' from the origin. When no such path exists, the origin is almost surely surrounded by a cycle of pairwise-intersecting needles. That is,

(12.3) $$\mathbb{P}_{\mu,l}(E) \begin{cases} < 1 & \text{if } l < l_c, \\ = 1 & \text{if } l > l_c, \end{cases}$$

where E is the event that there exists a component C of needles such that the origin of $\mathbb{R}^2$ lies in a bounded component of $\mathbb{R}^2 \setminus C$, and $\mathbb{P}_{\mu,l}$ denotes the probability measure governing the configuration of mirrors.

The needle percolation problem is a type of continuum percolation model, cf. the space–time percolation process of Section 6.6. Continuum percolation, and in particular the needle (or 'stick') model, has been summarized in [188, Sect. 8.5].

We return to the above Lorentz problem. Suppose that the photon is projected from the origin at angle θ to the x-axis, for given $\theta \in [0, 2\pi)$. Let

Θ be the set of all θ such that the trajectory of the photon is unbounded. It is clear from Theorem 12.2 that $\mathbb{P}_{\mu,l}(\Theta = \varnothing) = 1$ if $l > l_c$. The strength of the following theorem of Matthew Harris lies in the converse statement.

12.4 Theorem [134]. *Let μ be non-degenerate, with support a subset of the rational angles $\pi\mathbb{Q}$.*
  (a) *If $l > l_c$, then $\mathbb{P}_{\mu,l}(\Theta = \varnothing) = 1$.*
  (b) *If $l < l_c$, then*

$$\mathbb{P}_{\mu,l}(\Theta \text{ has Lebesgue measure } 2\pi) = 1 - \mathbb{P}_{\mu,l}(E) > 0.$$

That is to say, almost surely on the complement of E, the set Θ differs from the entire interval $[0, 2\pi)$ by a null set. The proof uses a type of dimension-reduction method, and utilizes a theorem concerning so-called 'interval-exchange transformations' taken from ergodic theory, see [149]. It is a key assumption for this argument that μ be supported within the rational angles.

Let $\eta(l) = \eta_\mu(l)$ be the probability that the light ray is bounded, having started by heading northwards from the origin. As above, $\eta_\mu(l) = 1$ when $l > l_c(\mu)$. In contrast, it is not known for general μ whether or not $\eta_\mu(l) < 1$ for sufficiently small positive l. It seems reasonable to conjecture the following. For any probability measure μ on $[0, \pi)$, there exists $l_r \in (0, l_c]$ such that $\eta_\mu(l) < 1$ whenever $l < l_r$. This conjecture is open even for the arguably most natural case when μ is uniform on $[0, \pi)$.

12.5 Conjecture. *Let μ be the uniform probability measure on $[0, \pi)$, and let l_c denote the critical length for the associated needle percolation problem (as in Theorem 12.2).*
  (a) *There exists $l_r \in (0, l_c]$ such that*

$$\eta(l) \begin{cases} < 1 & \text{if } l < l_r, \\ = 1 & \text{if } l > l_r, \end{cases}$$

  (b) *We have that $l_r = l_c$.*

As a first step, we seek a proof that $\eta(l) < 1$ for sufficiently small positive values of l. It is typical of such mirror problems that we lack even a proof that $\eta(l)$ is monotone in l.

12.4 Exercises

12.1 There are two ways of putting in the barriers in the percolation proof of Theorem 12.1, depending on whether one uses the odd or the even vertices. Use this fact to establish bounds for the tail of the size of the trajectory when the density of mirrors is 1.

12.2 In a variant of the square Lorentz lattice gas, NE mirrors occur with probability $\eta \in (0, 1)$ and NW mirrors otherwise. Show that the photon's trajectory is almost surely bounded.

12.3 Needles are dropped in the plane in the manner of a Poisson process with intensity 1. They have length l, and their angles to the horizontal are independent random variables with law μ. Show that there exists $l_c = l_c(\mu) \in (0, \infty]$ such that: the probability that the origin lies in an unbounded path of $\mathbb{R}^2$ intersecting no needle is strictly positive when $l < l_c$, and equals zero when $l > l_c$.

12.4 (continuation) Show that $l_c < \infty$ if and only if μ is non-degenerate.

References

Aaronson, J.

1. *An Introduction to Infinite Ergodic Theory*, American Mathematical Society, Providence, RI, 1997.

Ahlfors, L.

2. *Complex Analysis*, 3rd edn, McGraw-Hill, New York, 1979.

Aizenman, M.

3. Geometric analysis of ϕ^4 fields and Ising models, *Communications in Mathematical Physics* 86 (1982), 1–48.

4. The geometry of critical percolation and conformal invariance, *Proceedings STATPHYS 19 (Xiamen 1995)* (H. Bai-Lin, ed.), World Scientific, 1996, pp. 104–120.

5. Scaling limit for the incipient infinite clusters, *Mathematics of Multiscale Materials* (K. Golden, G. Grimmett, J. Richard, G. Milton, P. Sen, eds), IMA Volumes in Mathematics and its Applications, vol. 99, Springer, New York, 1998, pp. 1–24.

Aizenman, M., Barsky, D. J.

6. Sharpness of the phase transition in percolation models, *Communications in Mathematical Physics* 108 (1987), 489–526.

Aizenman, M., Barsky, D. J., Fernández, R.

7. The phase transition in a general class of Ising-type models is sharp, *Journal of Statistical Physics* 47 (1987), 343–374.

Aizenman, M., Burchard, A.

8. Hölder regularity and dimension bounds for random curves, *Duke Mathematical Journal* 99 (1999), 419–453.

9. Discontinuity of the magnetization in one-dimensional $1/|x-y|^2$ Ising and Potts models, *Journal of Statistical Physics* 50 (1988), 1–40.

Aizenman, M., Fernández, R.

10. On the critical behavior of the magnetization in high-dimensional Ising models, *Journal of Statistical Physics* 44 (1986), 393–454.

Aizenman, M., Grimmett, G. R.

11. Strict monotonicity for critical points in percolation and ferromagnetic models, *Journal of Statistical Physics* 63 (1991), 817–835.

Aizenman, M., Kesten, H., Newman, C. M.

12. Uniqueness of the infinite cluster and related results in percolation, *Percolation Theory and Ergodic Theory of Infinite Particle Systems* (H. Kesten, ed.), IMA Volumes in Mathematics and its Applications, vol. 8, Springer, New York, 1987, pp. 13–20.

Aizenman, M., Klein, A., Newman, C. M.

13. Percolation methods for disordered quantum Ising models, *Phase Transitions: Mathematics, Physics, Biology,* ... (R. Kotecký, ed.), World Scientific, Singapore, 1992, pp. 129–137.

Aizenman, M., Nachtergaele, B.

14. Geometric aspects of quantum spin systems, *Communications in Mathematical Physics* 164 (1994), 17–63.

Aizenman, M., Newman, C. M.

15. Tree graph inequalities and critical behavior in percolation models, *Journal of Statistical Physics* 36 (1984), 107–143.

Aldous, D. J.

16. Brownian excursions, critical random graphs and the multiplicative coalescent, *Annals of Probability* 25 (1997), 812–854.

17. The random walk construction of uniform spanning trees and uniform labelled trees, *SIAM Journal of Discrete Mathematics* 3 (1990), 450–465.

Aldous, D., Fill, J.

18. *Reversible Markov Chains and Random Walks on Graphs*, in preparation; http://www.stat.berkeley.edu/~aldous/RWG/book.html.

Alexander, K.

19. On weak mixing in lattice models, *Probability Theory and Related Fields* 110 (1998), 441–471.

20. Mixing properties and exponential decay for lattice systems in finite volumes, *Annals of Probability* 32 (2004), 441–487.

Alon, N., Spencer, J. H.

21. *The Probabilistic Method*, Wiley, New York, 2000.

Azuma, K.

22. Weighted sums of certain dependent random variables, *Tôhoku Mathematics Journal* 19 (1967), 357–367.

Bálint, A., Camia, F., Meester, R.

23. The high temperature Ising model on the triangular lattice is a critical percolation model, arXiv:0806.3020 (2008).

Barlow, R. N., Proschan, F.

24. *Mathematical Theory of Reliability*, Wiley, New York, 1965.

Baxter, R. J.

25. *Exactly Solved Models in Statistical Mechanics*, Academic Press, London, 1982.

Beckner, W.
26. Inequalities in Fourier analysis, *Annals of Mathematics* 102 (1975), 159–182.

Beffara, V.
27. Cardy's formula on the triangular lattice, the easy way, *Universality and Renormalization*, Fields Institute Communications, vol. 50, American Mathematical Society, Providence, RI, 2007, pp. 39–45.

Beffara, V., Duminil-Copin, H.
28. The self-dual point of the random-cluster model is critical for $q \geq 1$, in preparation (2010).

Benaïm, M., Rossignol, R.
29. Exponential concentration for first passage percolation through modified Poincaré inequalities, *Annales de l'Institut Henri Poincaré, Probabilités et Statistiques* 44 (2008), 544–573.

Ben-Or, M., Linial, N.
30. Collective coin flipping, *Randomness and Computation*, Academic Press, New York, 1990, pp. 91–115.

Benjamini, I., Kalai, G., Schramm, O.
31. First passage percolation has sublinear distance variance, *Annals of Probability* 31 (2003), 1970–1978.

Benjamini, I., Lyons, R., Peres, Y., Schramm, O.
32. Uniform spanning forests, *Annals of Probability* 29 (2001), 1–65.

Berg, J. van den
33. Disjoint occurrences of events: results and conjectures, *Particle Systems, Random Media and Large Deviations* (R. T. Durrett, ed.), Contemporary Mathematics no. 41, American Mathematical Society, Providence, R. I., 1985, pp. 357–361.
34. Approximate zero–one laws and sharpness of the percolation transition in a class of models including 2D Ising percolation, *Annals of Probability* 36 (2008), 1880–1903.

Berg, J. van den, Kesten, H.
35. Inequalities with applications to percolation and reliability, *Journal of Applied Probability* 22 (1985), 556–569.

Bezuidenhout, C. E., Grimmett, G. R.
36. The critical contact process dies out, *Annals of Probability* 18 (1990), 1462–1482.
37. Exponential decay for subcritical contact and percolation processes, *Annals of Probability* 19 (1991), 984–1009.
38. A central limit theorem for random walks in random labyrinths, *Annales de l'Institut Henri Poincaré, Probabilités et Statistiques* 35 (1999), 631–683.

Billingsley, P.
39. *Convergence of Probability Measures*, 2nd edn, Wiley, New York, 1999.

Björnberg, J. E.
40. Graphical representations of Ising and Potts models, Ph.D. thesis, Cambridge University (2009).

Björnberg, J. E., Grimmett, G. R.
41. The phase transition of the quantum Ising model is sharp, *Journal of Statistical Physics* 136 (2009), 231–273.

Bollobás, B.
42. The chromatic number of random graphs, *Combinatorica* 8 (1988), 49–55.
43. *Modern Graph Theory*, Springer, Berlin, 1998.
44. *Random Graphs*, 2nd edn, Cambridge University Press, Cambridge, 2001.

Bollobás, B., Grimmett, G. R., Janson, S.
45. The random-cluster process on the complete graph, *Probability Theory and Related Fields* 104 (1996), 283–317.

Bollobás, B., Riordan, O.
46. The critical probability for random Voronoi percolation in the plane is $1/2$, *Probability Theory and Related Fields* 136 (2006), 417–468.
47. A short proof of the Harris–Kesten theorem, *Bulletin of the London Mathematical Society* 38 (2006), 470–484.
48. *Percolation*, Cambridge University Press, Cambridge, 2006.

Bonami, A.
49. Étude des coefficients de Fourier des fonctions de $L^p(G)$, *Annales de l'Institut Fourier* 20 (1970), 335–402.

Borgs, C., Chayes, J. T., Randall, R.
50. The van-den-Berg–Kesten–Reimer inequality: a review, *Perplexing Problems in Probability* (M. Bramson, R. T. Durrett, eds), Birkhäuser, Boston, 1999, pp. 159–173.

Bourgain, J., Kahn, J., Kalai, G., Katznelson, Y., Linial, N.
51. The influence of variables in product spaces, *Israel Journal of Mathematics* 77 (1992), 55–64.

Broadbent, S. R., Hammersley, J. M.
52. Percolation processes I. Crystals and mazes, *Proceedings of the Cambridge Philosophical Society* 53 (1957), 629–641.

Broder, A. Z.
53. Generating random spanning trees, *30th IEEE Symposium on Foundations of Computer Science* (1989), 442–447.

Brook, D.
54. On the distinction between the conditional probability and joint probability approaches in the specification of nearest-neighbour systems, *Biometrika* 51 (1964), 481–483.

Burton, R. M., Keane, M.
55. Density and uniqueness in percolation, *Communications in Mathematical Physics* 121 (1989), 501–505.

230 References

Camia, F., Newman, C. M.
56. Continuum nonsimple loops and 2D critical percolation, *Journal of Statistical Physics* 116 (2004), 157–173.
57. Two-dimensional critical percolation: the full scaling limit, *Communications in Mathematical Physics* 268 (2006), 1–38.
58. Critical percolation exploration path and SLE_6: a proof of convergence, *Probability Theory and Related Fields* 139 (2007), 473–519.

Cardy, J.
59. Critical percolation in finite geometries, *Journal of Physics A: Mathematical and General* 25 (1992), L201.

Cerf, R.
60. The Wulff crystal in Ising and percolation models, *Ecole d'Eté de Probabilités de Saint Flour* XXXIV–2004 (J. Picard, ed.), Lecture Notes in Mathematics, vol. 1878, Springer, Berlin, 2006.

Cerf, R., Pisztora, Á.
61. On the Wulff crystal in the Ising model, *Annals of Probability* 28 (2000), 947–1017.
62. Phase coexistence in Ising, Potts and percolation models, *Annales de l'Institut Henri Poincaré, Probabilités et Statistiques* 37 (2001), 643–724.

Chayes, J. T., Chayes, L.
63. Percolation and random media, *Critical Phenomena, Random Systems and Gauge Theories* (K. Osterwalder, R. Stora, eds), Les Houches, Session XLIII, 1984, Elsevier, Amsterdam, 1986a, pp. 1001–1142.

Chayes, J. T., Chayes, L., Grimmett, G. R., Kesten, H., Schonmann, R. H.
64. The correlation length for the high density phase of Bernoulli percolation, *Annals of Probability* 17 (1989), 1277–1302.

Chayes, J. T., Chayes, L., Newman, C. M.
65. Bernoulli percolation above threshold: an invasion percolation analysis, *Annals of Probability* 15 (1987), 1272–1287.

Chayes, L., Lei, H. K.
66. Cardy's formula for certain models of the bond–triangular type, *Reviews in Mathematical Physics* 19 (2007), 511–565.

Chelkak, D., Smirnov, S.
67. Universality in the 2D Ising model and conformal invariance of fermionic observables, arXiv:0910.2045 (2009).

Clifford, P.
68. Markov random fields in statistics, *Disorder in Physical Systems* (G. R. Grimmett, D. J. A. Welsh, eds), Oxford University Press, Oxford, 1990, pp. 19–32.

Crawford, N., Ioffe, D.
69. Random current representation for transverse field Ising model, *Communications in Mathematical Physics* 296 (2010), 447–474.

Dobrushin, R. L.
  70. Gibbs state describing coexistence of phases for a three–dimensional Ising model, *Theory of Probability and its Applications* 18 (1972), 582–600.

Doeblin, W.
  71. Exposé de la théorie des chaînes simples constantes de Markoff à un nombre fini d'états, *Revue Mathématique de l'Union Interbalkanique* 2 (1938), 77–105.

Doyle, P. G., Snell, J. L.
  72. *Random Walks and Electric Networks*, Carus Mathematical Monographs, vol. 22, Mathematical Association of America, Washington, DC, 1984.

Dudley, R. M.
  73. *Real Analysis and Probability*, Wadsworth, Brooks & Cole, Pacific Grove CA, 1989.

Duminil-Copin, H., Smirnov, S.
  74. In preparation (2010).

Duplantier, B.
  75. Brownian motion, "diverse and undulating", *Einstein, 1905–2005, Poincaré Seminar 1 (2005)* (T. Damour, O. Darrigol, B. Duplantier, V. Rivasseau, eds), Progress in Mathematical Physics, vol. 47, Birkhäuser, Boston, 2006, pp. 201–293.

Durrett, R. T.
  76. On the growth of one-dimensional contact processes, *Annals of Probability* 8 (1980), 890–907.
  77. Oriented percolation in two dimensions, *Annals of Probability* 12 (1984), 999-1040.
  78. The contact process, 1974–1989, *Mathematics of Random Media* (W. E. Kohler, B. S. White, eds), American Mathematical Society, Providence, R. I., 1992, pp. 1–18.
  79. *Random Graph Dynamics*, Cambridge University Press, Cambridge, 2007.

Durrett, R., Schonmann, R. H.
  80. Stochastic growth models, *Percolation Theory and Ergodic Theory of Infinite Particle Systems* (H. Kesten, ed.), Springer, New York, 1987, pp. 85–119.

Ehrenfest, P.
  81. *Collected Scientific Papers* (M. J. Klein, ed.), North-Holland, Amsterdam, 1959.

Erdős, P., Rényi, A.
  82. The evolution of random graphs, *Magyar Tud. Akad. Mat. Kutató Int. Közl.* 5 (1960), 17–61.

Ethier, S., Kurtz, T.
  83. *Markov Processes: Characterization and Convergence*, Wiley, New York, 1986.

Falik, D., Samorodnitsky, A.

84. Edge-isoperimetric inequalities and influences, *Combinatorics, Probability, Computing* 16 (2007), 693–712.

Feder, T., Mihail, M.

85. Balanced matroids, *Proceedings of the 24th ACM Symposium on the Theory of Computing*, ACM, New York, 1992, pp. 26–38.

Ferrari, P. L., Spohn, H.

86. Scaling limit for the space–time covariance of the stationary totally asymmetric simple exclusion process, *Communications in Mathematical Physics* 265 (2006), 1–44.

Fortuin, C. M.

87. On the random-cluster model, Ph.D. thesis, University of Leiden (1971).

88. On the random-cluster model. II. The percolation model, *Physica* 58 (1972), 393–418.

89. On the random-cluster model. III. The simple random-cluster process, *Physica* 59 (1972), 545–570.

Fortuin, C. M., Kasteleyn, P. W.

90. On the random-cluster model. I. Introduction and relation to other models, *Physica* 57 (1972), 536–564.

Fortuin, C. M., Kasteleyn, P. W., Ginibre, J.

91. Correlation inequalities on some partially ordered sets, *Communications in Mathematical Physics* 22 (1971), 89–103.

Friedgut, E.

92. Influences in product spaces: KKL and BKKKL revisited, *Combinatorics, Probability, Computing* 13 (2004), 17–29.

Friedgut, E., Kalai, G.

93. Every monotone graph property has a sharp threshold, *Proceedings of the American Mathematical Society* 124 (1996), 2993–3002.

Georgii, H.-O.

94. *Gibbs Measures and Phase Transitions*, Walter de Gruyter, Berlin, 1988.

Gibbs, J. W.

95. *Elementary Principles in Statistical Mechanics*, Charles Scribner's Sons, New York, 1902; http://www.archive.org/details/elementary princi00gibbrich.

Gilbert, E. N.

96. Random graphs, *Annals of Mathematical Statistics* 30 (1959), 1141–1144.

Ginibre, J.

97. Reduced density matrices of the anisotropic Heisenberg model, *Communications in Mathematical Physics* 10 (1968), 140–154.

Glauber, R. J.

98. Time-dependent statistics of the Ising model, *Journal of Mathematical Physics* 4 (1963), 294–307.

Gowers, W. T.
99. How to lose your fear of tensor products, (2001); http://www.dpmms.cam.ac.uk/~wtg10/tensors3.html.

Graham, B. T., Grimmett, G. R.
100. Influence and sharp-threshold theorems for monotonic measures, *Annals of Probability* 34 (2006), 1726–1745.
101. Sharp thresholds for the random-cluster and Ising models, arXiv:0903.1501, *Annals of Applied Probability* (2010) (to appear).

Grassberger, P., Torre, A. de la
102. Reggeon field theory (Schögl's first model) on a lattice: Monte Carlo calculations of critical behaviour, *Annals of Physics* 122 (1979), 373–396.

Grimmett, G. R.
103. A theorem about random fields, *Bulletin of the London Mathematical Society* 5 (1973), 81–84.
104. The stochastic random-cluster process and the uniqueness of random-cluster measures, *Annals of Probability* 23 (1995), 1461–1510.
105. Percolation and disordered systems, *Ecole d'Eté de Probabilités de Saint Flour* XXVI–1996 (P. Bernard, ed.), Lecture Notes in Mathematics, vol. 1665, Springer, Berlin, 1997, pp. 153–300.
106. *Percolation*, 2nd edition, Springer, Berlin, 1999.
107. Stochastic pin-ball, *Random Walks and Discrete Potential Theory* (M. Picardello, W. Woess, eds), Cambridge University Press, Cambridge, 1999.
108. Infinite paths in randomly oriented lattices, *Random Structures and Algorithms* 18 (2001), 257–266.
109. *The Random-Cluster Model*, corrected reprint (2009), Springer, Berlin, 2006.
110. Space–time percolation, *In and Out of Equilibrium 2* (V. Sidoravicius, M. E. Vares, eds), Progress in Probability, vol. 60, Birkhäuser, Boston, 2008, pp. 305–320.
111. Three problems for the clairvoyant demon, *Probability and Mathematical Genetics* (N. H. Bingham, C. M. Goldie, eds), Cambridge University Press, Cambridge, 2010, pp. 379–395.

Grimmett, G. R., Hiemer, P.
112. Directed percolation and random walk, *In and Out of Equilibrium* (V. Sidoravicius, ed.), Progress in Probability, vol. 51, Birkhäuser, Boston, 2002, pp. 273–297.

Grimmett, G. R., Janson, S.
113. Random even graphs, Paper R46, *Electronic Journal of Combinatorics* 16 (2009).

Grimmett, G. R., Kesten, H., Zhang, Y.
114. Random walk on the infinite cluster of the percolation model, *Probability Theory and Related Fields* 96 (1993), 33–44.

Grimmett, G. R., Marstrand, J. M.

115. The supercritical phase of percolation is well behaved, *Proceedings of the Royal Society (London), Series A* 430 (1990), 439–457.

Grimmett, G. R., McDiarmid, C. J. H.

116. On colouring random graphs, *Mathematical Proceedings of the Cambridge Philosophical Society* 77 (1975), 313–324.

Grimmett, G. R., Menshikov, M. V., Volkov, S. E.

117. Random walks in random labyrinths, *Markov Processes and Related Fields* 2 (1996), 69–86.

Grimmett, G. R., Osborne, T. J., Scudo, P. F.

118. Entanglement in the quantum Ising model, *Journal of Statistical Physics* 131 (2008), 305–339.

Grimmett, G. R., Piza, M. S. T.

119. Decay of correlations in subcritical Potts and random-cluster models, *Communications in Mathematical Physics* 189 (1997), 465–480.

Grimmett, G. R., Stacey, A. M.

120. Critical probabilities for site and bond percolation models, *Annals of Probability* 26 (1998), 1788–1812.

Grimmett, G. R., Stirzaker, D. R.

121. *Probability and Random Processes*, 3rd edn, Oxford University Press, 2001.

Grimmett, G. R., Welsh, D. J. A.

122. *Probability, an Introduction*, Oxford University Press, Oxford, 1986.

123. John Michael Hammersley (1920–2004), *Biographical Memoirs of Fellows of the Royal Society* 53 (2007), 163–183.

Grimmett, G. R., Winkler, S. N.

124. Negative association in uniform forests and connected graphs, *Random Structures and Algorithms* 24 (2004), 444–460.

Gross, L.

125. Logarithmic Sobolev inequalities, *American Journal of Mathematics* 97 (1975), 1061–1083.

Halmos, P. R.

126. *Measure Theory*, Springer, Berlin, 1974.

127. *Finite-Dimensional Vector Spaces*, 2nd edn, Springer, New York, 1987.

Hammersley, J. M.

128. Percolation processes. Lower bounds for the critical probability, *Annals of Mathematical Statistics* 28 (1957), 790–795.

Hammersley, J. M., Clifford, P.

129. Markov fields on finite graphs and lattices, unpublished (1971); http://www.statslab.cam.ac.uk/~grg/books/hammfest/hamm-cliff.pdf.

Hammersley, J. M., Morton, W.

130. Poor man's Monte Carlo, *Journal of the Royal Statistical Society* (B) 16 (1954), 23–38.

Hammersley, J. M., Welsh, D. J. A.

131. First-passage percolation, subadditive processes, stochastic networks and generalized renewal theory, *Bernoulli, Bayes, Laplace Anniversary Volume* (J. Neyman, L. LeCam, eds), Springer, Berlin, 1965, pp. 61–110.

Hara, T., Slade, G.

132. Mean-field critical behaviour for percolation in high dimensions, *Communications in Mathematical Physics* 128 (1990), 333–391.

133. Mean-field behaviour and the lace expansion, *Probability and Phase Transition* (G. R. Grimmett, ed.), Kluwer, Dordrecht, 1994, pp. 87–122.

Harris, M.

134. Nontrivial phase transition in a continuum mirror model, *Journal of Theoretical Probability* 14 (2001), 299-317.

Harris, T. E.

135. A lower bound for the critical probability in a certain percolation process, *Proceedings of the Cambridge Philosophical Society* 56 (1960), 13–20.

136. Contact interactions on a lattice, *Annals of Probability* 2 (1974), 969–988.

Higuchi, Y.

137. A sharp transition for the two-dimensional Ising percolation, *Probability Theory and Related Fields* 97 (1993), 489–514.

Hintermann, A., Kunz, H., Wu, F. Y.

138. Exact results for the Potts model in two dimensions, *Journal of Statistical Physics* 19 (1978), 623–632.

Hoeffding, W.

139. Probability inequalities for sums of bounded random variables, *Journal of the American Statistical Association* 58 (1963), 13–30.

Holley, R.

140. Remarks on the FKG inequalities, *Communications in Mathematical Physics* 36 (1974), 227–231.

Hughes, B. D.

141. *Random Walks and Random Environments*, Volume I, Random Walks, Oxford University Press, Oxford, 1996.

Ising, E.

142. Beitrag zur Theorie des Ferromagnetismus, *Zeitschrift für Physik* 31 (1925), 253–258.

Janson, S., Knuth, D., Łuczak, T., Pittel, B.

143. The birth of the giant component, *Random Structures and Algorithms* 4 (1993), 233–358.

Janson, S., Łuczak, T., Ruciński, A.

144. *Random Graphs*, Wiley, New York, 2000.

Kahn, J., Kalai, G., Linial, N.
145. The influence of variables on Boolean functions, *Proceedings of 29th Symposium on the Foundations of Computer Science*, Computer Science Press, 1988, pp. 68–80.

Kahn, J., Neiman, M.
146. Negative correlation and log-concavity, arXiv:0712.3507, *Random Structures and Algorithms* (2010).

Kalai, G., Safra, S.
147. Threshold phenomena and influence, *Computational Complexity and Statistical Physics* (A. G. Percus, G. Istrate, C. Moore, eds), Oxford University Press, New York, 2006.

Kasteleyn, P. W., Fortuin, C. M.
148. Phase transitions in lattice systems with random local properties, *Journal of the Physical Society of Japan* 26 (1969), 11–14, Supplement.

Keane, M.
149. Interval exchange transformations, *Mathematische Zeitschrift* 141 (1975), 25–31.

Keller, N.
150. On the influences of variables on Boolean functions in product spaces, arXiv:0905.4216 (2009).

Kesten, H.
151. The critical probability of bond percolation on the square lattice equals $\frac{1}{2}$, *Communications in Mathematical Physics* 74 (1980a), 41–59.
152. *Percolation Theory for Mathematicians*, Birkhäuser, Boston, 1982.

Kirchhoff, G.
153. Über die Auflösung der Gleichungen, auf welche man bei der Untersuchung der linearen Verteilung galvanischer Strome gefuhrt wird, *Annalen der Physik und Chemie* 72 (1847), 497–508.

Klein, A.
154. Extinction of contact and percolation processes in a random environment, *Annals of Probability* 22 (1994), 1227–1251.
155. Multiscale analysis in disordered systems: percolation and contact process in random environment, *Disorder in Physical Systems* (G. R. Grimmett, ed.), Kluwer, Dordrecht, 1994, pp. 139–152.

Kotecký, R., Shlosman, S.
156. First order phase transitions in large entropy lattice systems, *Communications in Mathematical Physics* 83 (1982), 493–515.

Laanait, L., Messager, A., Miracle-Solé, S., Ruiz, J., Shlosman, S.
157. Interfaces in the Potts model I: Pirogov–Sinai theory of the Fortuin–Kasteleyn representation, *Communications in Mathematical Physics* 140 (1991), 81–91.

Langlands, R., Pouliot, P., Saint-Aubin, Y.

158. Conformal invariance in two-dimensional percolation, *Bulletin of the American Mathematical Society* 30 (1994), 1–61.

Lauritzen, S.

159. *Graphical Models*, Oxford University Press, Oxford, 1996.

Lawler, G.

160. *Conformally Invariant Processes in the Plane*, American Mathematical Society, Providence, RI, 2005.

Lawler, G. F., Schramm, O., Werner, W.

161. The dimension of the planar Brownian frontier is $4/3$, *Mathematics Research Letters* 8 (2001), 401–411.

162. Values of Brownian intersection exponents III: two-sided exponents, *Annales de l'Institut Henri Poincaré, Probabilités et Statistiques* 38 (2002), 109–123.

163. One-arm exponent for critical 2D percolation, *Electronic Journal of Probability* 7 (2002), Paper 2.

164. Conformal invariance of planar loop-erased random walks and uniform spanning trees, *Annals of Probability* 32 (2004), 939–995.

Levin, D. A., Peres, Y., Wilmer, E. L.

165. *Markov Chains and Mixing Times*, AMS, Providence, R. I., 2009.

Lieb, E., Schultz, T., Mattis, D.

166. Two soluble models of an antiferromagnetic chain, *Annals of Physics* 16 (1961), 407–466.

Liggett, T. M.

167. *Interacting Particle Systems*, Springer, Berlin, 1985.

168. Multiple transition points for the contact process on the binary tree, *Annals of Probability* 24 (1996), 1675–1710.

169. *Stochastic Interacting Systems: Contact, Voter and Exclusion Processes*, Springer, Berlin, 1999.

170. Interacting particle systems – an introduction, *ICTP Lecture Notes Series*, vol. 17, 2004; `http://publications.ictp.it/lns/vol17/vol17 toc.html`.

Lima, B. N. B. de

171. A note about the truncation question in percolation of words, *Bulletin of the Brazilian Mathematical Society* 39 (2008), 183–189.

Lima, B. N. B. de, Sanchis, R., Silva, R. W. C.

172. Percolation of words on $\mathbb{Z}^d$ with long range connections, arXiv:0905.4615 (2009).

Lindvall, T.

173. *Lectures on the Coupling Method*, Wiley, New York, 1992.

Linusson, S.

174. On percolation and the bunkbed conjecture, arXiv:0811.0949, *Combinatorics, Probability, Computing* (2010).

175. A note on correlations in randomly oriented graphs, arXiv:0905.2881 (2009).

Lorentz, H. A.
176. The motion of electrons in metallic bodies, I, II, III, *Koninklijke Akademie van Wetenschappen te Amsterdam, Section of Sciences* 7 (1905), 438–453, 585–593, 684–691.

Löwner, K.
177. Untersuchungen über schlichte konforme Abbildungen des Einheitskreises, I, *Mathematische Annalen* 89 (1923), 103–121.

Lubetzky, E., Sly, A.
178. Cutoff for the Ising model on the lattice, arXiv:0909.4320 (2009).
179. Critical Ising on the square lattice mixes in polynomial time, arXiv:1001. 1613 (2010).

Lyons, R.
180. Phase transitions on nonamenable graphs, *Journal of Mathematical Physics* 41 (2001), 1099–1126.

Lyons, R., Peres, Y.
181. *Probability on Trees and Networks*, in preparation, 2010; http://mypage. iu.edu/~rdlyons/prbtree/prbtree.html.

Lyons, T. J.
182. A simple criterion for transience of a reversible Markov chain, *Annals of Probability* 11 (1983), 393–402.

Madras, N., Slade, G.
183. *The Self-Avoiding Walk*, Birkhäuser, Boston, 1993.

Margulis, G.
184. Probabilistic characteristics of graphs with large connectivity, *Problemy Peredachi Informatsii* (in Russian) 10 (1974), 101–108.

Martinelli, F.
185. Lectures on Glauber dynamics for discrete spin models, *Ecole d'Eté de Probabilités de Saint Flour* XXVII–1997 (P. Bernard, ed.), Lecture Notes in Mathematics, vol. 1717, Springer, Berlin, pp. 93–191.

McDiarmid, C. J. H.
186. On the method of bounded differences, *Surveys in Combinatorics, 1989* (J. Siemons, ed.), LMS Lecture Notes Series 141, Cambridge University Press, Cambridge, 1989.
187. On the chromatic number of random graphs, *Random Structures and Algorithms* 1 (1990), 435–442.

Meester, R., Roy, R.
188. *Continuum Percolation*, Cambridge University Press, Cambridge, 1996.

Menshikov, M. V.
189. Coincidence of critical points in percolation problems, *Soviet Mathematics Doklady* 33 (1987), 856–859.

Menshikov, M. V., Molchanov, S. A., Sidorenko, A. F.

190. Percolation theory and some applications, *Itogi Nauki i Techniki* (Series of Probability Theory, Mathematical Statistics, Theoretical Cybernetics) 24 (1986), 53–110.

Moussouris, J.

191. Gibbs and Markov random fields with constraints, *Journal of Statistical Physics* 10 (1974), 11–33.

Nachtergaele, B.

192. A stochastic geometric approach to quantum spin systems, *Probability and Phase Transition* (G. R. Grimmett, ed.), Kluwer, Dordrecht, 1994, pp. 237–246.

Onsager, L.

193. Crystal statistics, I. A two-dimensional model with an order–disorder transition, *The Physical Review* 65 (1944), 117–149.

Peierls, R.

194. On Ising's model of ferromagnetism, *Proceedings of the Cambridge Philosophical Society* 36 (1936), 477–481.

Pemantle, R.

195. Choosing a spanning tree for the infinite lattice uniformly, *Annals of Probability* 19 (1991), 1559–1574.

196. The contact process on trees, *Annals of Probability* 20 (1992), 2089–2116.

197. Uniform random spanning trees, *Topics in Contemporary Probability and its Applications* (J. L. Snell, ed.), CRC Press, Boca Raton, 1994, pp. 1–54.

198. Towards a theory of negative dependence, *Journal of Mathematical Physics* 41 (2000), 1371–1390.

Petersen, K.

199. *Ergodic Theory*, Cambridge University Press, Cambridge, 1983.

Pólya, G.

200. Über eine Aufgabe betreffend die Irrfahrt im Strassennetz, *Mathematische Annalen* 84 (1921), 149–160.

201. Two incidents, *Collected Papers* (G. Pólya, G.-C. Rota, eds), vol. IV, The MIT Press, Cambridge, Massachusetts, 1984, pp. 582–585.

Potts, R. B.

202. Some generalized order–disorder transformations, *Proceedings of the Cambridge Philosophical Society* 48 (1952), 106–109.

Propp, D., Wilson, D. B.

203. How to get a perfectly random sample from a generic Markov chain and generate a random spanning tree of a directed graph, *Journal of Algebra* 27 (1998), 170–217.

Quas, A.

204. Infinite paths in a Lorentz lattice gas model, *Probability Theory and Related Fields* 114 (1999), 229–244.

Ráth, B.
205. Conformal invariance of critical percolation on the triangular lattice, Diploma thesis, http://www.math.bme.hu/~rathb/rbperko.pdf (2005).

Reimer, D.
206. Proof of the van den Berg–Kesten conjecture, *Combinatorics, Probability, Computing* 9 (2000), 27–32.

Rohde, S., Schramm, O.
207. Basic properties of SLE, *Annals of Mathematics* 161 (2005), 879–920.

Rossignol, R.
208. Threshold for monotone symmetric properties through a logarithmic Sobolev inequality, *Annals of Probability* 34 (2005), 1707–1725.
209. Threshold phenomena on product spaces: BKKKL revisited (once more), *Electronic Communications in Probability* 13 (2008), 35–44.

Rudin, W.
210. *Real and Complex Analysis*, 3rd edn, McGraw-Hill, New York, 1986.

Russo, L.
211. A note on percolation, *Zeitschrift für Wahrscheinlichkeitstheorie und Verwandte Gebiete* 43 (1978), 39–48.
212. On the critical percolation probabilities, *Zeitschrift für Wahrscheinlichkeitstheorie und Verwandte Gebiete* 56 (1981), 229–237.
213. An approximate zero–one law, *Zeitschrift für Wahrscheinlichkeitstheorie und Verwandte Gebiete* 61 (1982), 129–139.

Schonmann, R. H.
214. Metastability and the Ising model, *Documenta Mathematica,* Extra volume (Proceedings of the 1998 ICM) III (1998), 173–181.

Schramm, O.
215. Scaling limits of loop-erased walks and uniform spanning trees, *Israel Journal of Mathematics* 118 (2000), 221–288.
216. Conformally invariant scaling limits: an overview and collection of open problems, *Proceedings of the International Congress of Mathematicians, Madrid* (M. Sanz-Solé *et al.*, eds), vol. I, European Mathematical Society, Zurich, 2007, pp. 513–544.

Schramm, O., Sheffield, S.
217. Harmonic explorer and its convergence to SLE_4, *Annals of Probability* 33 (2005), 2127–2148.
218. Contour lines of the two-dimensional discrete Gaussian free field, *Acta Mathematica* 202 (2009), 21–137.

Schulman, L. S.
219. *Techniques and Applications of Path Integration*, Wiley, New York, 1981.

Seppäläinen, T.
220. Entropy for translation-invariant random-cluster measures, *Annals of Probability* 26 (1998), 1139–1178.

Seymour, P. D., Welsh, D. J. A.

221. Percolation probabilities on the square lattice, *Advances in Graph Theory* (B. Bollobás, ed.), Annals of Discrete Mathematics 3, North-Holland, Amsterdam, 1978, pp. 227–245.

Smirnov, S.

222. Critical percolation in the plane: conformal invariance, Cardy's formula, scaling limits, *Comptes Rendus des Séances de l'Académie des Sciences. Série I. Mathématique* 333 (2001), 239–244.

223. Critical percolation in the plane. I. Conformal invariance and Cardy's formula. II. Continuum scaling limit, http://www.math.kth.se/~stas/papers/ (2001).

224. Towards conformal invariance of 2D lattice models, *Proceedings of the International Congress of Mathematicians, Madrid, 2006* (M. Sanz-Solé *et al.*, eds), vol. II, European Mathematical Society, Zurich, 2007, pp. 1421–1452.

225. Conformal invariance in random cluster models. I. Holomorphic fermions in the Ising model, arXiv:0708.0039, *Annals of Mathematics* (2010).

Smirnov, S., Werner, W.

226. Critical exponents for two-dimensional percolation, *Mathematics Research Letters* 8 (2001), 729–744.

Strassen, V.

227. The existence of probability measures with given marginals, *Annals of Mathematical Statistics* 36 (1965), 423–439.

Swendsen, R. H., Wang, J. S.

228. Nonuniversal critical dynamics in Monte Carlo simulations, *Physical Review Letters* 58 (1987), 86–88.

Szász, D.

229. *Hard Ball Systems and the Lorentz Gas*, Encyclopaedia of Mathematical Sciences, vol. 101, Springer, Berlin, 2000.

Talagrand, M.

230. Isoperimetry, logarithmic Sobolev inequalities on the discrete cube, and Margulis' graph connectivity theorem, *Geometric and Functional Analysis* 3 (1993), 295–314.

231. On Russo's approximate zero–one law, *Annals of Probability* 22 (1994), 1576–1587.

232. On boundaries and influences, *Combinatorica* 17 (1997), 275–285.

233. On influence and concentration, *Israel Journal of Mathematics* 111 (1999), 275–284.

Tóth, B.

234. Persistent random walks in random environment, *Probability Theory and Related Fields* 71 (1986), 615–625.

Werner, W.

235. Random planar curves and Schramm–Loewner evolutions, *Ecole d'Eté de Probabilités de Saint Flour* XXXII–2002 (J. Picard, ed.), Springer, Berlin, 2004, pp. 107–195.

236. Lectures on two-dimensional critical percolation, *Statistical Mechanics* (S. Sheffield, T. Spencer, eds), IAS/Park City Mathematics Series, Vol. 16, AMS, Providence, R. I., 2009, pp. 297–360.

237. *Percolation et Modéle d'Ising*, Cours Spécialisés, vol. 16, Société Mathématique de France, 2009.

Wierman, J. C.

238. Bond percolation on the honeycomb and triangular lattices, *Advances in Applied Probability* 13 (1981), 298–313.

Williams, G. T., Bjerknes, R.

239. Stochastic model for abnormal clone spread through epithelial basal layer, *Nature* 236 (1972), 19–21.

Wilson, D. B.

240. Generating random spanning trees more quickly than the cover time, *Proceedings of the 28th ACM on the Theory of Computing*, ACM, New York, 1996, pp. 296–303.

Wood, De Volson

241. Problem 5, *American Mathematical Monthly* 1 (1894), 99, 211–212.

Wu, F. Y.

242. The Potts model, *Reviews in Modern Physics* 54 (1982), 235–268.

Wulff, G.

243. Zur Frage der Geschwindigkeit des Wachsturms und der Auflösung der Krystallflächen, *Zeitschrift für Krystallographie und Mineralogie* 34 (1901), 449–530.

Index